Rakesh Kumar and Geeta Arora

Fuzzy Clustering Algorithms

De Gruyter Series on the Applications of Mathematics in Engineering and Information Sciences

Edited by
Mangey Ram

Volume 23

Rakesh Kumar and Geeta Arora

Fuzzy Clustering Algorithms

Symbiotic Relationship Between Mathematics and Similarity Measures

DE GRUYTER

Authors
Prof. Dr Rakesh Kumar
Department of Mathematics
Lovely Professional University
Phagwara 144411
Punjab
India
E-mail: rakeshmalhan23@gmail.com

Prof. Dr Geeta Arora
Department of Mathematics
Lovely Professional University
Phagwara 144411
Punjab
India
E-mail: geetadma@gmail.com

ISBN 978-3-11-149238-4
e-ISBN (PDF) 978-3-11-149427-2
e-ISBN (EPUB) 978-3-11-149576-7
ISSN 2626-5427

Library of Congress Control Number: 2025944967

Bibliographic information published by the Deutsche Nationalbibliothek
The Deutsche Nationalbibliothek lists this publication in the Deutsche Nationalbibliografie; detailed bibliographic data are available on the internet at http://dnb.dnb.de.

Cover image: MF3d/E+/Getty Images
Typesetting: Integra Software Services Pvt. Ltd.
Printing and binding: CPI books GmbH, Leck

www.degruyterbrill.com
Questions about General Product Safety Regulation:
productsafety@degruyterbrill.com

MIX
Papier | Fördert gute Waldnutzung
FSC® C083411

Dedicated
to our cherished families, whose boundless love, steadfast support, and endless encouragement have been the foundation of our endeavors,
and
to the Almighty, whose divine guidance and blessings have inspired and sustained us, we dedicate this work with deep gratitude and reverence.

Preface

In the rapidly evolving field of data science, the ability to uncover patterns and structures within complex datasets is paramount. Clustering, a cornerstone of unsupervised learning, provides a powerful framework for grouping data points based on their similarities, offering insights across diverse domains such as market research, image segmentation, bioinformatics, and beyond. This book delves into the fascinating world of clustering, with a particular focus on two advanced methodologies: Fuzzy *C*-means (FCM) and possibilistic fuzzy *C*-means (PFCM) algorithms.

The journey begins with an exploration of clustering fundamentals, introducing key concepts, definitions, and distance measures that form the foundation of clustering techniques. From there, we dive into the nuances of FCM, a soft clustering approach that allows data points to belong to multiple clusters with varying degrees of membership, offering a flexible and nuanced perspective on data organization. Building on this, we introduce PFCM, a robust hybrid algorithm that combines the strengths of fuzzy and possibilistic approaches to address challenges such as noise sensitivity and overlapping clusters.

This book is designed for students, researchers, and practitioners in data science, machine learning, and related fields who seek a comprehensive understanding of these clustering techniques. Through detailed explanations, mathematical derivations, practical examples, and programming implementations in MATLAB, R, and Python, we aim to provide both theoretical depth and hands-on applicability. Real-world case studies, such as clustering IPL players' performance and optimizing warehouse locations, illustrate the practical power of these methods, while problem sets encourage readers to apply their knowledge to new scenarios.

Our goal is to equip readers with the tools to harness clustering algorithms effectively, enabling them to uncover hidden patterns in data and make informed decisions in their respective fields. Whether you are a novice exploring data analysis or an experienced professional seeking advanced techniques, this book serves as a guide to mastering FCM and PFCM, paving the way for innovative solutions in an increasingly data-driven world.

We hope this exploration inspires curiosity and empowers you to push the boundaries of what's possible with clustering.

https://doi.org/10.1515/9783111494272-202

Contents

List of figures

https://doi.org/10.1515/9783111494272-204

List of tables

https://doi.org/10.1515/9783111494272-205

About the authors

Dr Rakesh Kumar is doctorate in mathematics from Lovely Professional University. He has authored and coauthored over 47 research papers published in reputed international journals and conferences. His contributions span fuzzy logic applications, actuarial modeling, multiobjective optimization, clustering algorithms, smart city management enabled by AI, and computational approaches for data science and engineering problems. His work demonstrates a strong interdisciplinary approach, bridging mathematical theory and real-world applications.

Dr Geeta Arora earned her Ph.D. from IIT Roorkee, India, in 2011. She is working as a professor in the Department of Mathematics at Lovely Professional University, Punjab, India. She has over 12 years of teaching experience and published around 68 research papers (Scopus indexed) in both international and national journals. She has authored 15 book chapters in national and international publications. Also, she has written a book on Vedic mathematics and a book on essential statistics. She has also edited a book on numerical methods in Taylor & Francis and IGI Global, and is currently working on books with Elsevier and Taylor & Francis. Her area of research is the development of numerical methods and statistics. She received the research appreciation award in 2017, 2019, and 2024 from Lovely Professional University, Punjab, India. She has conducted several workshops on MATLAB and Vedic mathematics for students and faculty members within and outside the university campus. She has supervised nine Ph.D. scholars and is currently guiding six Ph.D. scholars. She has published eight books with national publisher along with five ongoing books with national and international publishers.

https://doi.org/10.1515/9783111494272-206

Chapter 1
Introduction

1.1 Clustering

Clustering is a data analysis technique frequently employed in strategy creation, market research, and business system design. The partitioning of commodities is a standard concern in inventory control and management. In many industries, diverse materials and components of machinery or other apparatus must be handled to attain the intended objective. To enhance the effectiveness of material management, a recommended strategy is to categorize various materials into categories [1–4]. Clustering facilitates the identification of normative thresholds within the data. It is a subfield of interdisciplinary analysis and unsupervised learning for pattern recognition. Clusters are predominantly utilized throughout several scientific disciplines, including machine learning, pattern recognition, optimization, image segmentation, taxonomy, medicine, geology, and engineering [5, 6].

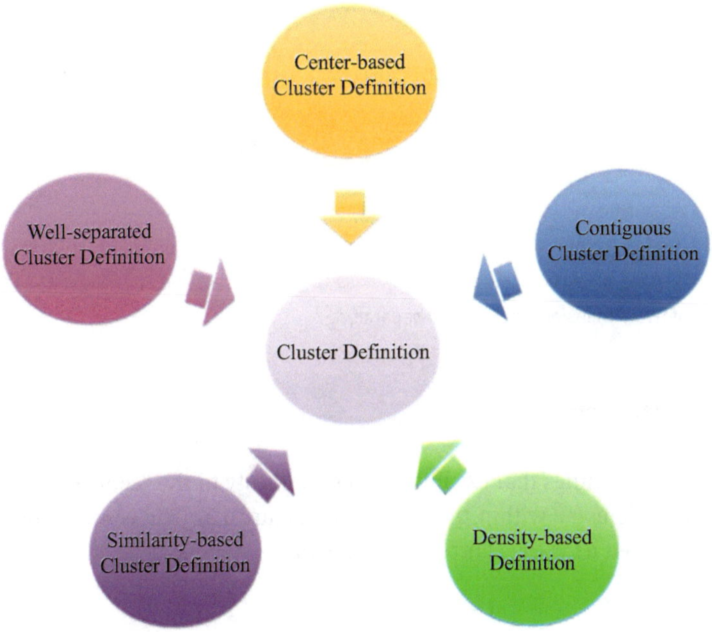

Figure 1.1: Different types of clustering.

https://doi.org/10.1515/9783111494272-001

Clustering is a compilation of data attributes that are relatively proximate in feature values or distinct from certain groupings. Clustering is an unsupervised learning technique used to identify groups of similar data objects by matching their features. As several definitions have been proposed in the past, there are a plethora of definitions to pick from. These definitions are predicated on terms such as same and alike, or on a specific type of cluster [2, 7–10]. The definitions are most of an ambiguous, circular sort. This illustrates the challenge of establishing a universally accepted meaning for the term "cluster". Several prevalent working definitions of a cluster are presented, with the most significant ones mentioned below (see Figure 1.1).

1.1.1 Well-separated cluster definition

A cluster is a set of points that exhibit more similarity to a particular point inside the cluster than to any point external to it. A threshold is often used to classify points (see Figure 1.2) in a cluster that are relatively similar (or close) to one another. Nonetheless, points in the peripheries of clusters exhibit more similarity to characteristics in other clusters than to those within their own cluster.

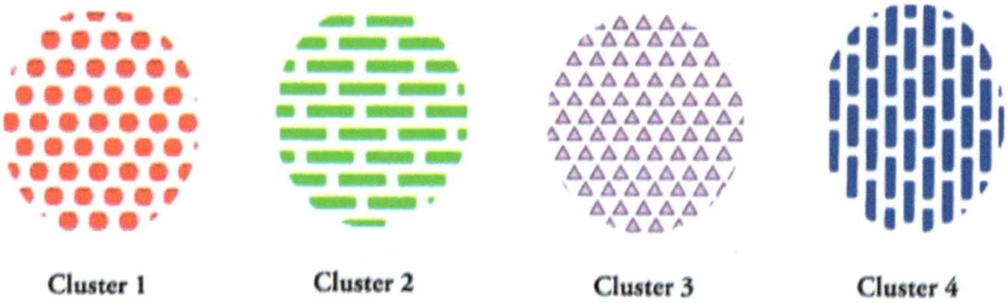

Figure 1.2: Four clusters of two-dimensional points well-separated.

1.1.2 Centre-based cluster definition

Clusters are assemblages of things that exhibit greater similarity to the "centre" of the cluster than to any other cluster within the group (see Figure 1.3). The centroid of a cluster is invariably situated near the centre of the cluster.

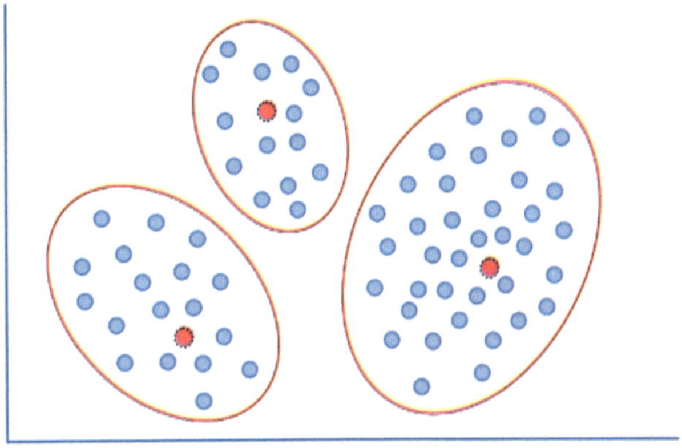

Figure 1.3: Centre-based cluster.

1.1.3 Contiguous cluster definition

A cluster is a collection of points where the closest point to any given point is not a member of the cluster (see Figure 1.4).

Figure 1.4: Contiguous cluster.

1.1.4 Density-based cluster definition

Unsupervised learning approaches that identify recognizable classes or clusters (see Figure 1.5) in the data are known as density-based clustering. The notion that clusters within a data space represents the regions of high point density, separated from other clusters by regions of low point density [11, 12]. The delineated areas with low point density typically indicate noise or outliers.

Figure 1.5: Density-based cluster.

1.1.5 Similarity-based cluster definition

A cluster is a collection of analogous things, while elements in different clusters are unlike (see Figure 1.6). A cluster may be defined as a collection of points that collectively constitute a region characterized by a uniform specified attribute, such as density or structure.

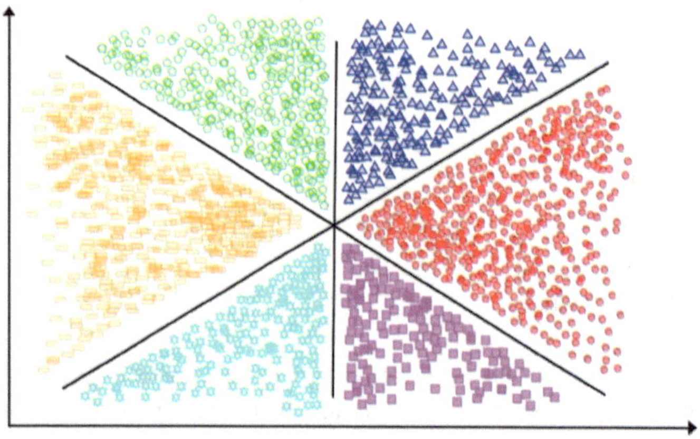

Figure 1.6: Similarity-based cluster.

1.2 Distance measures

The objective of calculating the distance between clusters is essential for distinguishing them within the data collection. Each distance metric possesses distinct advantages; however, none is universally applicable to all clustering challenges. The concept of a metric space is essential and crucial in mathematics. Diverse metric functions provide unique metric spaces [13, 14]. In pattern recognition and machine learning, the measurement of distance and the contrast between sample pairs are very important [15, 16]. Utilizing advanced numerical optimization, we may get distinguishing features through a logical representation of the distance function and determine if two samples belong to the same class. From this viewpoint, distance metric learning and dimensionality reduction aim to acquire high-level semantic distances, wherein identical input items are mapped to proximate spots, while disparate objects are distinguished from one another. Distance metrics have been efficiently used for vast scenarios, including image selection, visual monitoring, and prototype classification [17, 18].

1.2.1 Minkowski distance

If X be a universal space and $X_{i\,k}$ and $X_{j\,k} \in X$, then the Minkowski metric distance on crisp data is defined as

$$D_{w(i,j)_k} = \left[\sum_{k=1}^{n} \left| X_{i\,k} - X_{j\,k} \right|^{w} \right]^{\frac{1}{w}} \tag{1.1}$$

The Minkowski's measure holds for $w \in [1, \infty)$. For the special case of $w = 1$, it becomes Hamming distance, and when $w = 2$, it is Euclidean distance.

1.2.2 Mahalanobis distance

It is measured by deducting the Euclidean distance between two points through their standard deviation, which is expressed as

$$M_{d(i,j)_k} = \left\{ \left[(X_i - X_j)^T V^{-1} (X_i - X_j) \right] \right\}^{1/2} \tag{1.2}$$

where V is the sample covariance matrix. If the covariance matrix V is the identity matrix, then the $M_{d(i,j)}$ reduces to the Euclidean distance. If V is diagonal, then the determined distance measure is called a normalized Euclidean distance [19] defined as

$$M_{d(i,j)_k} = \sqrt{\frac{\sum_{k=1}^{n} \left| X_{i\,k} - X_{j\,k} \right|^2}{V}}$$

1.2.3 Cosine distance

The cosine of angle is utilized to measure the resemblance between two vectors. It is described as [20]

$$D_{\cos(i,j)_k} = 1 - \frac{\sum_{k=1}^{n} X_{i\,k}.X_{j\,k}}{\sqrt{\sum_{k=1}^{n} (X_{i\,k})^2}.\sqrt{\sum_{k=1}^{n} (X_{j\,k})^2}} \tag{1.3}$$

If $X_{i\,k}.X_{j\,k} \geq 0$, then $D_{\cos(i,j)_k} \in [0,1]$.

1.2.4 Correlation distance

This distance measure is derived from the Spearman correlation coefficient(s) and is defined as [21]

$$D_{\mathrm{corr}(i,j)_k} = 1 - \frac{\sum_{k=1}^{n} (X_{ik} - \overline{X_{ik}}).(X_{jk} - \overline{X_{jk}})}{\sqrt{\sum_{k=1}^{n} (X_{ik} - \overline{X_{ik}})^2}.\sqrt{\sum_{k=1}^{n} (X_{jk} - \overline{X_{jk}})^2}} \tag{1.4}$$

where $\overline{X_{ik}} = \frac{1}{n}\sum_{k=1}^{n} X_{i\,k}$ and $\overline{X_{jk}} = \frac{1}{n}\sum_{k=1}^{n} X_{jk}$.

1.2.5 Chebyshev distance

It measures the distance between the characteristics of a pair of data points. It is represented as

$$D_{\max(i,j)} = \max_{k} \left| X_{i\,k} - X_{j\,k} \right| \tag{1.5}$$

It would take less time to measure the distance between the datasets.

1.3 Types of clustering

Clustering algorithms are classified into two main groups: partitional and hierarchical.

1.3.1 Hierarchical clustering

Hierarchical methodologies produce a structured decomposition of data collection [22, 23]. A hierarchical clustering arranges smaller clusters as subdivisions of bigger clusters (see Figure 1.7). Each clustering level is hierarchically nested inside the hierarchical structure clustering. These clusters overlap and do not serve dynamism. The recent study of the data disaggregates the dataset into various subgroups. This cluster tree is referred to as a dendrogram and has clusters of differing sizes. Hierarchical clustering has two distinct kinds: agglomerative clustering and divisive clustering.

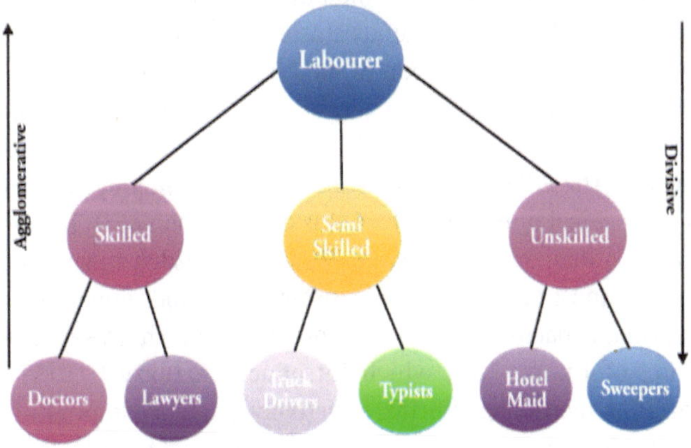

Figure 1.7: Hierarchical clustering.

1.3.1.1 Agglomerative clustering

The bottom-up approach is considered agglomerative clustering. It establishes an individual cluster for each object initially. Subsequently, related clusters are amalgamated into extended clusters until all particles are allocated to one or more clusters or are concluded.

1.3.1.2 Divisive clustering

The top-down method is also known as divisive clustering. In this case, each piece of data starts out in a cluster. The cluster is then broken up into smaller groups until each piece of data is in its own cluster, which creates a nested clustering.

1.3.2 Partitional clustering

Partitional clustering decomposes a set of disjoint clusters. To split a database into subspace segments, consider applying one or more clustering methods to the entire dataset. Each cluster must include a minimum of one item, ensuring that every item is encompassed inside a cluster. To identify a superior partitioning, the dataset is divided into k initial partitions, followed by several rounds to enhance the cluster quality. Objects are transported across clusters during the iteration, enhancing clustering efficiency [24–26]. The algorithm stops when no item can be transferred and quality cannot be enhanced.

1.3.2.1 Type of the partitional clustering

There are two types of partitional clustering: (a) k-means clustering and (b) fuzzy c-means (FCM) clustering. Hard c-means (HCM) clustering, sometimes referred to as k-means clustering, is fundamentally a partitioning method that assigns each data point to exactly one cluster. Thus, the problem of ambiguous cluster boundaries may be addressed using fuzzy clustering. The distinct division of data in fuzzy clustering is enhanced by a less stringent partitioning criterion, wherein fuzzy connections denote the association among data points. Fuzzy clustering is characterized by complexity, ambiguity, and a lack of rigidity. The FCM technique is the most often employed fuzzy clustering approach in several critical domains [27]. FCM is an extension of HCM, and it has demonstrated improved outcomes compared to HCM [28]. Bezdek's proposed FCM algorithms are perhaps the most often employed clustering methods due to their steady uncertainty attributes and capacity to retain far more detail than hard segmentation. FCM is an expansion of Dunn's [29], which introduced a fuzzy ISODATA method. The conventional FCM method performs effectively with predominantly noiseless data, but it is vulnerable to noisy prototypes due to its atypical feature data.

1.3.2.2 Hard *c*-means clustering

HCM clustering is the principal and most prevalent research methodology for hard clusters. A configuration list of overlaps constitutes the cluster assignment, denoted as $\mho = \{\mho_1, \mho_2, \ldots, \mho_K\}$, where $\mho_i$ is the defining index vector of samples for the ith cluster. Define $L\ (y_i)$ as the y_i cluster. The objective function of the HCM is equivalent to the summation of the squared errors of the fitness function, and it can be readily structured to minimize the Euclidean distances between pairs of points inside the same cluster, that is

$$min\ J_{HCM}(SSE) = \sum_{k=1}^{N} \sum_{j=1}^{c} ||y_k - v_j||^2 \tag{1.6}$$

Subject to

$$v_j = \frac{\sum_{k=1}^{n} y_{jk}}{n} \tag{1.7}$$

The notation $y_k, v_j \in \lambda_k$ indicates that both the y_k and v_j data points are allocated to the kth cluster λ_k. While clustering can be effectively applied in k – means [30–32], it is simple to converge to the local optimum, and besides, the clustering assignment relies heavily on the clustering number K, which users have to pre-assign empirically.

Chapter 2
Fuzzy *C*-means clustering algorithms

2.1 Introduction

Fuzzy *C*-means (FCM) is also one of the most common soft clustering algorithms, particularly for its capacity to operate on overlaying data where objects or points lack to belong to one and only one cluster. However, unlike older "hard" clustering algorithm (such as *k*-means), where every data point is associated with a single cluster, FCM provides membership values of points with respect to each cluster in the range between 0 and 1. These membership values are the limit of the points of each cluster where we can take a more detailed impression of the data structure.

Technically, FCM attempts to minimize an objective function that is a weighted series of distances between the data points and the cluster centres by iteratively optimizing the cluster membership values and the cluster centres. The fuzziness index (m) regulates the extent to which the membership distribution is soft: the greater the values of m, the more memberships are distributed, and the smaller the values of m, the closer they are to crisp assignments. Owing to this flexibility, FCM finds its application in medical image segmentation, pattern recognition, and data mining - particularly where data inherently overlap in clustering.

In general, *C*-means models are the most advanced clustering point-prototype models. The FCM clustering methodology is an iterative clustering method that lowers the function jury item to generate the appropriate "*c*" divisions [33]. The standard FCM [27, 34] framework is as follows: fuzzy clustering models, as opposed to crisp *c*-partitions, allow components to correspond to several clusters to varying degrees in a fuzzy situation. Requirements specified in formulas may include normalizing the total membership of each data function vector to one, implying that the variable cannot fully belong to many more clusters than it now does. The fuzzifier "m" must be greater than one, because the optimization problem is a clear case when $m = 1$. In the literature, the most commonly used value for the fuzzifier is $m = 2$.

Let's discuss the difference between smooth and FCM using a favourite illustration from the research. It can be proved that the fuzzy case of clustering provides a far better explanation of the needle point because it assigns membership approximately evenly to each cluster. Fuzzy clusters, also known as smooth clustering or soft *k*-means, worry about what happens to more than one cluster. Clusters are inconsistent because clustering, also known as cluster evaluation, requires distributing information among clusters in a way that makes items in the same cluster almost identical.

Essentially, FCM is an overprovisioning algorithm. In terms of image analysis, the operation space will be divided into the clusters that the consumer requests. It is unaffected by the number of clusters that exist in the data collection. It is primarily concerned with the probabilistic constraint on either the membership interest or the member interest.

https://doi.org/10.1515/9783111494272-002

The acquired membership value, together with the use of FCM, extends the pixel's "degree of sharing" deep into the function space of the multiple clusters. In other words, it seems to represent the degree to which clusters will share a pixel [35, 36].

The FCM technique was initially suggested by Bezdek in 1981 as a modification of the hard *C*-means approach. The FCM method is a free-of-supervision clustering technique that addresses a wide variety of problems involving statistical analysis of features, aggregation, and classifier development. FCM is extensively utilized in many disciplines, including engineering for agriculture, astronomy, chemistry, geological science, and the analysis of images, medical evaluation, form analysis, and the identification of targets. The utilization of this technique for analysis is achieved by the utilization of the segregation of various input data points. The construction of collections depends on the measurement of separation between data points, after which the central point of every group is determined. FCM is a popular clustering of information approach that attempts to divide given information into a specified number of categories, denoted as "*n*". Each cluster is distinguished by a robust interconnection with the remaining clusters and demonstrates a significant level of membership. In contrast, data points that are located significantly distant from the centroid of a cluster exhibit a diminished level of membership to that particular cluster [37, 38].

2.2 FCM

It is within the context of this minimization problem that the FCM clustering algorithm is defined. The FCM technique is defined by optimizing the following objective function:

$$J_{\mathrm{FCM}} = \sum_{j=1}^{N} \sum_{i=1}^{C} \mu_{ij}^{m} \left\| x_j - v_i \right\|^2 ; \sum_{k=1}^{C} \mu_{kj} = 1 \tag{2.1}$$

In this academic context, "*N*" represents the total data points and "*p*" represents the number of variables or features. Here, "*m*" is a parameter denoting the degree of fuzziness, typically set to 2. Lastly, the distance measure between x_j and v_i is represented by $\left\| x_j - v_i \right\|^2$:

$$v_i = \frac{\sum_{j=1}^{N} \mu_{ij}^{m} x_j}{\sum_{j=1}^{N} \mu_{ij}^{m}} \tag{2.2}$$

$$\mu_{ij} = \frac{1}{\sum_{k=1}^{C} \left(\frac{\|x_j - v_i\|^2}{\|x_j - v_k\|^2} \right)^{\frac{1}{m-1}}} \tag{2.3}$$

This algorithm commences by initializing the algorithm with a randomly generated partition matrix $U_{N\times C}$. The cluster centres, denoted as v_i, and the membership degrees, denoted as μ_{ij}, are iteratively updated using the equations mentioned above until converges. In the above algorithm, the termination condition is defined as the absolute difference $\left\|v^{t+1} - v^t\right\|$ being less than ε, where ε is a positive value. In light of this important development, we give the MATLAB algorithmic code for the FCM clustering technique below. The code for almost every algorithm will function identically.

Algorithm 1. Algorithms for fuzzy *C*-means.

Input:

The data samples
Parameter m, ε
Number of cluster centers C

Output:

Cluster prototypes v_i with optimum membership values μ_{ij}

Begin algorithm

Step 1: Take initial cluster centers
a. Normalize data
b. Calculate initial cluster centers using appropriate method
Step 2: Calculate membership values
a. Calculate distance between x_j and v_i using $\| x_j - v_i \|^2$
b. Calculate μ_{ij} using (3)
Step 3: Update v_i
a. Calculate cluster centers v_i using equation (2).
Step 4: Terminating condition
a. Check if the termination condition for v_i using $\left|v_i^{(t)} - v_i^{(t-1)}\right| < \varepsilon$ is satisfied.
If yes then terminate algorithm.
If no then go to step 2.

End algorithm

FCM:

```
m=2;
Choose number of cluster 'C' center convenient for your data;
Initially take random C×N membership value matrix;
Start "For" loop for number of iterations (say r);
Take data set;
Find initial cluster centers using formula;
```

$$v_i = \frac{\sum_{j=1}^{N} \mu_{ij}^m x_j}{\sum_{j=1}^{N} \mu_{ij}^m}$$

```
In next step, find distance 'd' between cluster centers and data points using Euclidean distance formula;
```

Again start "For" loop for finding membership values iteratively as many times as number of data points.
Find value of μ_{ij} using;

$$\mu_{ij} = \frac{1}{\sum_{k=1}^{C} \left(\frac{||x_j - v_i||^2}{||x_j - v_k||^2} \right)^{\frac{1}{m-1}}}$$

End the loop for membership values;
In next step put the terminating condition for checking optimization of solution.
End the "For" loop for number of iterations 'r';
In next step find the value of objective function;

$$J_{FCM} = \sum_{j=1}^{N} \sum_{i=1}^{C} \mu_{ij}^{m} \, ||x_j - v_i||^2$$

And plot the data points and centers on the graph for better analysis of data.

2.3 R code

```
# Load necessary library
library(e1071)
# Import your dataset
df <- read.csv("path_to_your_file.csv")
# Perform Fuzzy C-Means Clustering
set.seed(123) # For reproducibility
fcm_model <- cmeans(df, centers = 3, iter.max = 100, m = 2)
# Display clustering results
print(fcm_model)
# Cluster membership matrix (degree of belonging to each cluster)
membership <- fcm_model$membership
print(membership)
# Add cluster labels to the original data
df$Cluster <- as.factor(fcm_model$cluster)
# Visualize the clusters
library(ggplot2)
ggplot(df, aes(x = df[,1], y = df[,2], color = Cluster)) +
 geom_point(size = 3) +
 labs(title = "Fuzzy C-Means Clustering", x = "Feature 1", y =
"Feature 2")
```

2.4 Python code

```
import numpy as np
import pandas as pd
from skfuzzy import cmeans
import matplotlib.pyplot as plt

# Load your dataset
df = pd.read_csv("path_to_your_file.csv")

data = df.iloc[:, :2].values # Use relevant features for clustering

# Perform Fuzzy C-Means Clustering
cntr, u, u0, d, jm, p, fpc = cmeans(
   data.T, # Transpose data for skfuzzy format
   c=3, # Number of clusters
   m=2, # Fuzziness parameter
   error=0.005,
   maxiter=1000,
   init=None
)

# Assign cluster labels
cluster_labels = np.argmax(u, axis=0)
df['Cluster'] = cluster_labels

# Visualize the clusters
plt.figure(figsize=(8, 6))
for i in range(3):
    plt.scatter(data[cluster_labels == i, 0], data[cluster_labels == i,
1], label=f'Cluster {i+1}')
plt.title('Fuzzy C-Means Clustering')
plt.xlabel('Feature 1')
plt.ylabel('Feature 2')
plt.legend()
plt.show()
```

2.5 Solved examples

Statement of Problem 1

The study investigates 15 CSK IPL players' stats including batting averages, over rates, and selection positions. Table 2.1 provides all the required information related to IPL players. Employing the FCM clustering technique, we want to categorize the players into three unique clusters. Our analysis groups players into batters, all-rounders and fast bowlers. Through this technique, we discover the strongest bowling all-rounder performers within our dataset.

Table 2.1: CSK team members' batting and bowling statistics along with their early designations.

Player	Bat_Avg	Bowl_Avg	Initial class
MM Ali	24.96	22.78	Batsman
S Anirudha	19	50	Bowler
R Ashwin	9.5	24.22	Bowler
S Badrinath	30.65	50	Batsman
L Balaji	3.66	26.09	Bowler
DE Bollinger	21	18.72	Batsman
DJ Bravo	25.1	22.47	Batsman
DL Chahar	10.83	28.2	Bowler
SM Curran	22	30.54	Batsman
MS Dhoni	40.53	50	Batsman
F du Plessis	35.33	50	Batsman
RD Gaikwad	39.36	50	Batsman
MS Gony	12.5	30.73	Bowler
Harbhajan Singh	7.5	25.3	Bowler

Research inquiries:

(i) **Classification of players:**
What is the use of FCM clustering for categorizing players into three distinct categories based on their batting and bowling averages: Our system identifies players into three separate groups, which are batsmen, bowling all-rounders, and bowlers.

(ii) **Determination of optimal bowling all-rounders:**
What are the profiles of players in our sample who perform well both as bowlers and batsmen, and how do they stack up versus other player types?

(iii) **Cluster analysis:**
How does each performance segment appear different through its bowling and batting statistics?

The discovered clusters show how much they match the player categories from step one.

(iv) **Performance analysis:**
By using FCM clustering, we can find player performance patterns that we cannot see through simple data review.

Solution:
Iteration 1:
Step 1: Consider Bat_Avg = X1 and Bowl_Avg = X2.

X1	24.96	19	9.5	30.65	3.66	21	25.1	10.83	22	40.53	35.33	39.36	12.5	7.5
X2	22.78	50	24.22	50	26.09	18.72	22.47	28.2	30.54	50	50	50	30.73	25.3

Step 2: Starting partition of the membership function into three clustering centres.

Mu1	0.009	0.233	0.954	0.019	0.914	0.113	0.012	0.972	0.213	0.019	0.000	0.012	0.851	0.976
Mu2	0.003	0.566	0.006	0.957	0.017	0.020	0.004	0.004	0.070	0.953	0.996	0.969	0.025	0.004
MU3	0.988	0.200	0.040	0.024	0.068	0.867	0.984	0.024	0.717	0.029	0.000	0.019	0.125	0.020

Step 3: Take $m = 2$.

Mu1^2	Mu2^2	Mu3^2	X1*Mu1^2	X2*Mu1^2	X1*Mu2^2	X2*Mu2^2	X1*Mu3^2	X2*Mu3^2
8.91E-05	9.32E-06	0.975172	0.002224	0.002029637	0.000233	0.000212	24.34029	22.21441982
0.05444	0.320807	0.040111	1.034363	2.722008372	6.095334	16.04035	0.762111	2.005554055
0.909592	3.58E-05	0.001623	8.641121	22.03031173	0.00034	0.000868	0.01542	0.039312236
0.00035	0.91563	0.000596	0.010713	0.01747694	28.06405	45.78148	0.018276	0.029813494
0.83604	0.000306	0.004645	3.059906	21.8122833	0.00112	0.007985	0.017	0.121184326
0.012696	0.000417	0.75152	0.266622	0.237674597	0.008756	0.007806	15.78191	14.06844884
0.000152	1.62E-05	0.96755	0.003817	0.003417342	0.000407	0.000364	24.28549	21.74083743
0.944111	1.74E-05	0.000585	10.22472	26.62392766	0.000188	0.00049	0.006331	0.01648642
0.045331	0.004912	0.514096	0.997284	1.384412102	0.108059	0.150005	11.31012	15.70050067
0.000353	0.907604	0.000814	0.014316	0.017660613	36.78521	45.38022	0.032974	0.040678988
0	0.991394	0	0	0	35.02595	49.56971	0	0
0.000154	0.938844	0.000348	0.006059	0.00769727	36.95291	46.94221	0.013694	0.017396339
0.723552	0.000611	0.01554	9.044404	22.23476191	0.007638	0.018778	0.194256	0.47755799
0.951905	1.5E-05	0.000419	7.139289	24.08320183	0.000112	0.000379	0.003144	0.010607031
4.478766	**4.080619**	**3.273019**	**40.44484**	**121.1768633**	**143.0503**	**203.9009**	**76.78103**	**76.48279764**

Step 4: Value of cluster centers

v1	9.030355	27.05586
v2	35.05603	49.96812
v3	23.45878	23.36766

Step 5: Distances of data points from all cluster centers

D1	272.04	625.83	8.26	993.84	29.77	212.76	279.26	4.55	180.35	1518.66	1218.10	1446.32	25.54	5.43
D2	841.12	257.80	1316.08	19.41	1555.88	1174.02	855.27	1060.75	547.91	29.97	0.08	18.53	878.88	1367.85
D3	2.60	729.16	195.57	761.00	399.40	27.65	3.50	182.84	53.57	1000.71	850.21	962.13	174.30	258.42

Step 6: Final membership values

NMu1	0.009435	0.233324	0.953719	0.018692	0.914342959	0.112651	0.012326	0.971666	0.212958	0.018798	6.24E-05	0.012411	0.850648	0.975648
NMu2	0.003051	0.566417	0.005988	0.956896	0.017497082	0.020415	0.004025	0.004166	0.070097	0.952675	0.999848	0.968933	0.024717	0.00387
NMu3	0.987514	0.200258	0.040293	0.024412	0.068159959	0.866934	0.983649	0.024169	0.716945	0.028527	8.95E-05	0.018656	0.124634	0.020482

Step 7: $e = 0.0001$.

e1	4.5E-06	2.1E-07	6.0E-06	3.6E-06	9.3E-06	2.7E-05	6.3E-06	1.2E-05	4.7E-05	3.7E-06	6.2E-05	3.1E-06	2.9E-05	8.0E-06
e2	1.5E-06	1.9E-05	9.9E-07	1.0E-05	2.5E-06	4.8E-06	2.1E-06	1.4E-06	1.3E-05	7.4E-06	4.2E-03	6.4E-06	2.5E-06	1.4E-06
e3	6.0E-06	1.9E-05	5.0E-06	6.8E-06	6.8E-06	3.1E-05	8.4E-06	1.0E-05	6.0E-05	3.7E-06	8.9E-05	3.4E-06	2.7E-05	6.5E-06

Statement of Problem 2

An organization has warehouses for life-saving drugs at six different locations, whose coordinates are given in the Table 2.2. The organization is in the process of locating one more critical central warehouse, which will distribute drugs to all the existing warehouses on emergency requests. Find the number of optimum locations of the new facility (warehouse) based on the FCM clustering concept.

Table 2.2: Thirty warehouses whose coordinates are summarized.

Existing warehouse number	Coordinates of centroids
1	(2.5,3.5)
2	(2.7,3.6)
3	(2.2,3.2)
4	(3.0,3.0)
5	(2.7,3.1)
6	(2.3,3.5)
7	(3.1,3.0)
8	(3.3,2.9)
9	(3.0,3.5)
10	(2.9, 2.9)
11	(6.0,7.2)
12	(6.5,7.0)
13	(5.9,6.9)
14	(6.3,7.3)
15	(5.9,7.0)
16	(6.8,7.0)
17	(6.5,6.9)
18	(6.2,7.1)
19	(7.2,7.2)
20	(6.6,7.1)
21	(11.1,13.5)
22	(12.2,12.5)
23	(11.5,11.9)
24	(13.2,13.3)
25	(11.5,12.5)
26	(12.0,15.0)
27	(13.2,13.2)
28	(11.2,11.9)
29	(12.9,13.7)
30	(13.9,13.0)

Solution:

The data are shown in Table 2.2. Assume that we want to determine a fuzzy pseudo-partition with three clusters (i.e. $c = 3$). Assume further that we choose $m = 2$. $||\,.\,||$ is the Euclidean distance, and the initial fuzzy pseudo-partition is $U^{(0)} = [U_1, U_2, U_3]$. Table 2.3–2.8 shows the respective results obtained at different iterations.

Table 2.3: The membership function for fuzzy clustering.

Input data		Membership function for fuzzy clustering		
X_1	X_2	U_1	U_2	U_3
2.5	3.5	.9	.05	.05
2.7	3.6	.89	.055	.055
2.2	3.2	.88	.06	.06
3.0	3.0	.87	.065	.065
2.7	3.1	.86	.07	.07
2.3	3.5	.84	.09	.07
3.1	3.0	.81	.11	.08
3.3	2.9	.8	.1	.1
3.0	3.5	.79	.11	.1
2.9	2.9	.75	.15	.1
6.0	7.2	.05	.9	.05
6.5	7.0	.055	.89	.055
5.9	6.9	.06	.88	.06
6.3	7.3	.065	.87	.065
5.9	7.0	.07	.86	.07
6.8	7.0	.09	.84	.07
6.5	6.9	.11	.81	.08
6.2	7.1	.1	.8	.1
7.2	7.2	.11	.79	.1
6.6	7.1	.15	.75	.1
11.1	13.5	.05	.05	.9
12.2	12.5	.055	.055	.89
11.5	11.9	.06	.06	.88
13.2	13.3	.065	.065	.87
11.5	12.5	.07	.07	.86
12.0	15.0	.07	.09	.84
13.2	13.2	.08	.11	.81
11.2	11.9	.1	.1	.8
12.9	13.7	.1	.11	.79
13.9	13.0	.1	.15	.75

Iteration 1:

Table 2.4: Cluster centres *V* and the corresponding membership function values of matrix *U* (Iteration 1).

Cluster centres		V_1	V_2	V_3
		(2.8723,3.3594)	**(6.3981,7.0939)**	**(12.0856,12.9097)**
Average values		**(2.87,3.36)**	**(6.40, 7.09)**	**12.09,12.91)**
Data points		U_1	U_2	U_3
2.5	3.5	.9935	.0056	.0009
2.7	3.6	.9961	.0034	.0005
2.2	3.2	.9832	.0143	.0024
3.0	3.0	.9941	.0051	.0008
2.7	3.1	.9962	.0033	.0005
2.3	3.5	.9866	.0115	.0019
3.1	3.0	.9925	.0065	.001
3.3	2.9	.9836	.0143	.0022
3.0	3.5	.9983	.0015	.0002
2.9	2.9	.9918	.007	.0011
6.0	7.2	.0069	.9907	.0024
6.5	7.0	.0007	.999	.0003
5.9	6.9	.0129	.9833	.0038
6.3	7.3	.0019	.9973	.0008
5.9	7.0	.0113	.9853	.0035
6.8	7.0	.0059	.9914	.0027
6.5	6.9	.0019	.9974	.0007
6.2	7.1	.0016	.9979	.0006
7.2	7.2	.019	.9698	.0112
6.6	7.1	.0015	.9979	.0006
11.1	13.5	.0075	.0203	.9722
12.2	12.5	.0011	.0029	.9961
11.5	11.9	.0089	.0267	.9643
13.2	13.3	.0066	.0161	.9773
11.5	12.5	.0032	.0091	.9877
12.0	15.0	.0188	.0437	.9375
13.2	13.2	.0064	.0155	.9781
11.2	11.9	.0121	.0372	.9508
12.9	13.7	.0061	.0147	.9792
13.9	13.0	.0146	.0344	.9509

Iteration 2:

Table 2.5: Cluster centres V and the corresponding membership function values of matrix U (Iteration 2).

Cluster centres		V_1	V_2	V_3
		(2.7721,3.2222)	**(6.3917,7.0736)**	**(12.2709,13.0401)**
Average values		**(2.77,3.22)**	**(6.39,7.07)**	**(12.27,13.04)**
Data points		U_1	U_2	U_3
2.5	3.5	.9938	.0054	.0008
2.7	3.6	.9935	.0057	.0008
2.2	3.2	.9884	.0099	.0016
3.0	3.0	.9959	.0036	.0005
2.7	3.1	.9992	.0007	.0001
2.3	3.5	.9884	.01	.0016
3.1	3.0	.9935	.0057	.0008
3.3	2.9	.984	.014	.0021
3.0	3.5	.994	.0053	.0007
2.9	2.9	.9953	.004	.0006
6.0	7.2	.0064	.9913	.0023
6.5	7.0	.0006	.9991	.0002
5.9	6.9	.0115	.9851	.0034
6.3	7.3	.002	.9971	.0009
5.9	7.0	.0101	.9867	.0032
6.8	7.0	.0056	.9918	.0026
6.5	6.9	.0015	.9979	.0006
6.2	7.1	.0014	.9981	.0005
7.2	7.2	.0183	.9708	.0109
6.6	7.1	.0015	.9979	.0007
11.1	13.5	.0087	.0241	.9671
12.2	12.5	.0017	.0047	.9936
11.5	11.9	.0119	.0365	.9516
13.2	13.3	.0044	.0108	.9849
11.5	12.5	.0053	.0156	.979
12.0	15.0	.0165	.0392	.9443
13.2	13.2	.0042	.0104	.9854
11.2	11.9	.0156	.0493	.9351
12.9	13.7	.0039	.0095	.9866
13.9	13.0	.0116	.0279	.9605

Iteration 3:

Table 2.6: Cluster centres *V* and the corresponding membership function values of matrix *U* (Iteration 3).

Cluster centres		V_1	V_2	V_3
		(2.7713,3.2212)	**(6.3918,7.0738)**	**(12.2855,13.0517)**
Average values		**(2.77,3.22)**	**(6.39,7.07)**	**(12.29,13.05)**
Data points		U_1	U_2	U_3
2.5	3.5	.9938	.0054	.0008
2.7	3.6	.9934	.0057	.0008
2.2	3.2	.9885	.0099	.0016
3.0	3.0	.9959	.0036	.0005
2.7	3.1	.9992	.0007	.0001
2.3	3.5	.9884	.01	.0016
3.1	3.0	.9935	.0057	.0008
3.3	2.9	.984	.014	.002
3.0	3.5	.9939	.0053	.0007
2.9	2.9	.9954	.004	.0006
6.0	7.2	.0064	.9913	.0023
6.5	7.0	.0006	.9991	.0002
5.9	6.9	.0115	.9851	.0034
6.3	7.3	.002	.9971	.0009
5.9	7.0	.0101	.9867	.0032
6.8	7.0	.0056	.9918	.0026
6.5	6.9	.0015	.9979	.0006
6.2	7.1	.0014	.9981	.0005
7.2	7.2	.0183	.9709	.0108
6.6	7.1	.0015	.9979	.0006
11.1	13.5	.0089	.0245	.9667
12.2	12.5	.0018	.0049	.9933
11.5	11.9	.0122	.0374	.9504
13.2	13.3	.0042	.0104	.9854
11.5	12.5	.0056	.0162	.9782
12.0	15.0	.0164	.0389	.9448
13.2	13.2	.0041	.0101	.9859
11.2	11.9	.016	.0504	.9336
12.9	13.7	.0037	.0091	.9872
13.9	13.0	.0114	.0274	.9612

Iteration 4:

Table 2.7: Cluster centres V and the corresponding membership function values of matrix U (Iteration 4).

Cluster centres		V_1	V_2	V_3
		(2.7713,3.2212)	**(6.3919,7.0739)**	**(12.2868,13.0527)**
Average values		**(2.77,3.22)**	**(6.39,7.07)**	**(12.69,13.05)**
Data points		U_1	U_2	U_3
2.5	3.5	.9938	.0054	.0008
2.7	3.6	.9934	.0057	.0008
2.2	3.2	.9885	.0099	.0016
3.0	3.0	.9959	.0036	.0005
2.7	3.1	.9992	.0007	.0001
2.3	3.5	.9884	.01	.0016
3.1	3.0	.9935	.0057	.0008
3.3	2.9	.984	.014	.002
3.0	3.5	.9939	.0053	.0007
2.9	2.9	.9954	.004	.0006
6.0	7.2	.0064	.9913	.0023
6.5	7.0	.0006	.9991	.0002
5.9	6.9	.0115	.9851	.0034
6.3	7.3	.002	.9971	.0009
5.9	7.0	.0101	.9867	.0032
6.8	7.0	.0056	.9919	.0026
6.5	6.9	.0015	.9979	.0006
6.2	7.1	.0014	.9981	.0005
7.2	7.2	.0183	.9709	.0108
6.6	7.1	.0015	.9979	.0006
11.1	13.5	.0089	.0245	.9666
12.2	12.5	.0018	.0049	.9933
11.5	11.9	.0122	.0375	.9503
13.2	13.3	.0042	.0104	.9854
11.5	12.5	.0056	.0163	.9781
12.0	15.0	.0163	.0388	.9448
13.2	13.2	.004	.0101	.9859
11.2	11.9	.016	.0505	.9335
12.9	13.7	.0037	.0091	.9872
13.9	13.0	.0114	.0274	.9612

Iteration 5:

Table 2.8: Cluster centres *V* and the corresponding membership function values of matrix *U* (Iteration 5).

Cluster centres		V_1	V_2	V_3
		(2.7713,3.2212)	**(6.3919,7.07339)**	**(12.2869,13.0528)**
Average values		**(2.77,3.22)**	**(6.39,7.07)**	**(12.29,13.05)**
Data points		U_1	U_2	U_3
2.5	3.5	.9938	.0054	.0008
2.7	3.6	.9934	.0057	.0008
2.2	3.2	.9885	.0099	.0016
3.0	3.0	.9959	.0036	.0005
2.7	3.1	.9992	.0007	.0001
2.3	3.5	.9884.	.01	.0016
3.1	3.0	.9935	.0057	.0008
3.3	2.9	.984	.014	.002
3.0	3.5	.9939	.0053	.0007
2.9	2.9	.9954	.004	.0006
6.0	7.2	.0064	.9913	.0023
6.5	7.0	.0006	.9991	.0002
5.9	6.9	.0115	.9851	.0034
6.3	7.3	.002	.9971	.0009
5.9	7.0	.0101	.9867	.0032
6.8	7.0	.0056	.9919	.0026
6.5	6.9	.0015	.9979	.0006
6.2	7.1	.0014	.9981	.0005
7.2	7.2	.0183	.9709	.0108
6.6	7.1	.0015	.9979	.0006
11.1	13.5	.0089	.0245	.9666
12.2	12.5	.0018	.0049	.9933
11.5	11.9	.0122	.0375	.9503
13.2	13.3	.0042	.0104	.9854
11.5	12.5	.0056	.0163	.9781
12.0	15.0	.0163	.0388	.9448
13.2	13.2	.004	.0101	.9859
11.2	11.9	.016	.0505	.9335
12.9	13.7	.0037	.0091	.9872
13.9	13.0	.0114	.0274	.9612

Iteration 6:

Table 2.9: Cluster centres V and the corresponding membership function values of matrix U (Iteration 6).

Cluster centres		V_1	V_2	V_3
		(2.7713,3.2212)	(6.3919,7.0739)	(12.2869,13.0528)
Average values		**(2.77,3.22)**	**(6.39,7.07)**	**(12.29,13.05)**
Data points		U_1	U_2	U_3
2.5	3.5	.9938	.0054	.0008
2.7	3.6	.9934	.0057	.0008
2.2	3.2	.9885	.0099	.0016
3.0	3.0	.9959	.0036	.0005
2.7	3.1	.9992	.0007	.0001
2.3	3.5	.9884	.01	.0016
3.1	3.0	.9935	.0057	.0008
3.3	2.9	.984	.014	.002
3.0	3.5	.9939	.0053	.0007
2.9	2.9	.9954	.004	.0006
6.0	7.2	.0064	.9913	.0023
6.5	7.0	.0006	.9991	.0002
5.9	6.9	.0115	.9851	.0034
6.3	7.3	.002	.9971	.0009
5.9	7.0	.0101	.9867	.0032
6.8	7.0	.0056	.9919	.0026
6.5	6.9	.0015	.9979	.0006
6.2	7.1	.0014	.9981	.0005
7.2	7.2	.0183	.9709	.0108
6.6	7.1	.0015	.9979	.0006
11.1	13.5	.0089	.0245	.9666
12.2	12.5	.0018	.0049	.9933
11.5	11.9	.0122	.0375	.9503
13.2	13.3	.0042	.0104	.9854
11.5	12.5	.0056	.0163	.9781
12.0	15.0	.0163	.0388	.9448
13.2	13.2	.004	.0101	.9859
11.2	11.9	.016	.0505	.9335
12.9	13.7	.0037	.0091	.9872
13.9	13.0	.0114	.0274	.9612

Then, the algorithm stops for $t = 6$, because $\max\left\{\left| u_{ij}^{k+1} - u_{ij}^{k} \right|\right\} < 3.8545^{-7}$ and we obtain the pseudo-partition defined in Table 2.9. The three clusters centres are $V_1 = (2.77, 3.22)$, $V_2 = (6.39, 7.07)$, and $V_3 = (12.29, 13.05)$ and $J_{FCM} = 3.1394$.These are three practice exercises that resemble your FCM clustering problem, in increasing dimensions (2D, 3D, and 4D), with clear directions on what to classify and how to evaluate your cluster performance:

2.6 Practice set

Practice question 1: (2D) – football players' performance

Football players (2D)

Player	Passing accuracy	Tackles won
Player_1	87.79	82.05
Player_2	84.45	88.76
Player_3	79.92	56.92
Player_4	85.49	75.91
Player_5	76.73	33.49

Problem statement:
A set of data has performance measures of 20 football players. Each of the players is measured at:
- Passing accuracy (%)
- Percentage of tackles won (%)

Cluster players into three categories using FCM clustering:
- Playmakers (lots of passing but low tackling)
- Defenders (low passing, high tackling)
- All-rounders (good at all of them, in general)

Objectives:
1. Classify players in three fuzzy groups.
2. Identify those players who must succeed in tackling and passing.
3. Compare cluster centres so as to find the difference in the roles of players.
4. Reveal invisible patterns of performance that were missed by classification based on thresholds.

Practice question 2: (3D) – basketball player segmentation

Basketball players (3D)

Player	PPG	RPG	APG
Player_1	12.9	10.1	6.7
Player_2	18.2	6.3	3.2
Player_3	27.3	3.9	2.5
Player_4	10.5	13.4	8.6
Player_5	22.1	5.6	4.9

Problem statement:
You have the statistics of 18 basketball players as follows:
- Points per game (PPG)
- Rebounds per game (RPG)
- Assist per game (APG)

Put them into groups using FCM cluster into:
- Scorers
- Rebounders
- Playmakers

Objectives:
1. Group players depending on their main performance indicators.
2. Locate hybrid personnel (e.g. scoring playmakers).
3. Determine the characterization of cluster centroid of specialized and all-round performances.
4. View fuzzy membership values to determine those players holding overlapping categories.

Practice question 3: (4D) – e-sports gamers' role detection

E-sports players (4D)

Player	Kills	Assists	Deaths	Damage taken
Player_1	18.1	20.5	3.7	2,301.5
Player_2	5.3	12.4	6.2	4,205.3
Player_3	9.6	22.1	2.9	1,100.6
Player_4	14.7	15.3	10.3	3,954.8
Player_5	3.2	18.6	7.1	3,500.0

Problem statement:

You are evaluating 25 professional e-sports players of a MOBA game and the metrics are as follows:

Kills/match
- Helps per match
- Deaths/match
- Damage it takes per match

Core turns them over to FCM to cluster them into:
- Attackers
- Advocates
- Tanks

Objectives:
1. Determine role-related groups of the 4D performance domain.
2. Look at hybrid players who can tank, attack, and support.
3. Research the variation of cluster centres with regard to kill-assist-death-damage profiles.
4. Find subtle forms of performance that cannot be divided into easy categories.

Chapter 3
Possibilistic *C*-means

3.1 Introduction

A clustering algorithm, such as possibilistic *C*-means (PCM) has also been designed to enhance the robustness of the results obtained in the cluster. In contrast to the fuzzy *C*-means (FCM), where membership values are used and calculated in such a way that they have to add to one across clusters per data point, PCM uses typicality values in order to assess how typical each point of a set is of one cluster in particular.

By emphasizing typicality over forced membership distribution, PCM is no longer vulnerable to outliers and is able to determine clusters based on an individual data point's belongingness strength. Nevertheless, a challenge that is well-known based on PCM is that it has an inclination of overlapping clusters when several cluster centres crumble into one point. Nevertheless, PCM would still be beneficial in the scenarios in which noise-resistance is an absolute necessity, and where the association of points with clusters is of greater significance than the overlap between them.

PCM refers to a particular implementation of Krishnapuram and Keller's possibility theory [39, 40]. The methodology being referred to is a clustering method that involves the modification of the goal function of the FCM algorithm [24, 41]:

$$\mathrm{J_{PCM}} = \sum_{\mathrm{j=1}}^{\mathrm{N}} \sum_{\mathrm{i=1}}^{\mathrm{C}} (\mathrm{t_{ij}})^{\mathrm{m}} \left\| x_j - v_i \right\|^2 + \sum_{\mathrm{i=1}}^{\mathrm{C}} \mathrm{Y_i} \sum_{\mathrm{j=1}}^{\mathrm{N}} (1 - \mathrm{t_{ij}})^{\mathrm{m}}$$

where

$$\mathrm{Y_i} = \frac{\sum_{\mathrm{j=1}}^{\mathrm{N}} \mathrm{t_{ij}^m} \left\| x_j - v_i \right\|^2}{\sum_{\mathrm{j=1}}^{\mathrm{N}} \mathrm{t_{ij}^m}},$$

where

$$K = 1$$

$$\mathrm{t_{ij}} = \frac{1}{1 + \left(\frac{\left\| x_j - v_i \right\|^2}{\mathrm{Y_i}} \right)^{\frac{1}{\mathrm{m-1}}}}$$

and

$$v_i = \frac{\sum_{j=1}^{N} t_{ij}^m x_j}{\sum_{j=1}^{N} t_{ij}^m}$$

https://doi.org/10.1515/9783111494272-003

The variable t_{ij} is linked to the possibilistic terms of the PCM method. It is an element of the typicality matrix $T_{C\times N}$ and serves as an indicator of the data point's membership in that cluster [42]. The PCM algorithm commences by utilizing centres determined by the FCM algorithm, followed by the iterative updating of v_i and t_{ij} until convergence is achieved.

3.2 Mathematical derivation for PCM

Derivation for possibilistic membership t_{ij}

For finding possibilistic membership, compute the partial derivative of J_{PCM} with respect to t_{ij},

$$\frac{\partial J_{\mathrm{PCM}}}{\partial t_{ij}} = mt_{ij}^{m-1}\|x_j - v_i\|^2 - mY_i(1-t_{ij})^{m-1}$$

Set it to zero:

$$=> mt_{ij}^{m-1}\|x_j - v_i\|^2 - mY_i(1-t_{ij})^{m-1} = 0$$

$$=> mt_{ij}^{m-1}\|x_j - v_i\|^2 = mY_i(1-t_{ij})^{m-1}$$

$$=> \frac{t_{ij}^{m-1}}{(1-t_{ij})^{m-1}} = \frac{Y_i}{\|x_j - v_i\|^2}$$

Now solve it for t_{ij}:

$$=> \left(\frac{t_{ij}}{1-t_{ij}}\right)^{m-1} = \frac{Y_i}{\|x_j - v_i\|^2}$$

$$=> \frac{t_{ij}}{1-t_{ij}} = \left(\frac{Y_i}{\|x_j - v_i\|^2}\right)^{\frac{1}{m-1}}$$

$$=> t_{ij} = \frac{1}{1+\left(\frac{\|x_j - v_i\|^2}{Y_i}\right)^{\frac{1}{m-1}}}$$

Derivation for cluster centre v_i

Now, for finding cluster centre, we take the derivative of J_{PCM} with respect to v_i and then set it to zero:

$$\frac{\partial J_{\text{PCM}}}{\partial v_i} = \sum_{j=1}^{N} -2\left(t_{ij}\right)^m \left(x_j - v_i\right) = 0$$

$$=> \sum_{j=1}^{N} \left(t_{ij}\right)^m \left(x_j - v_i\right) = 0$$

Solving for v_i we get

$$v_i = \frac{\sum_{j=1}^{N} t_{ij}^m x_j}{\sum_{j=1}^{N} t_{ij}^m}$$

Derivation for scale parameter Y_i

Scale parameter comes from empirical optimization, not purely calculus. It can't be calculated as v_i and t_{ij} as t_{ij} depends on Y_i. It is usually estimated using weighted mean square error. So,

$$Y_i = \frac{\sum_{j=1}^{N} t_{ij}^m \left\|x_j - v_i\right\|^2}{\sum_{j=1}^{N} t_{ij}^m}$$

PCM:

---

m=2;
Choose number of cluster 'C' center convenient for your data;
Initially take random C×N typicality value matrix;
Start "For" loop for number of iterations (say r);
Take data set;
Find initial cluster centers using formula;

$$v_i = \frac{\sum_{j=1}^{N} t_{ij}^m x_j}{\sum_{j=1}^{N} t_{ij}^m},$$

Find distance 'd' between cluster centers and data points using Euclidean distance formula;
In next step find value of 'Y_i' using;

$$Y_i = K\frac{\sum_{j=1}^{N} t_{ij}^m \left\|x_j - v_i\right\|^2}{\sum_{j=1}^{N} t_{ij}^m}$$

Again start "For" loop for finding typicality values iteratively as many times as number of data points.
Find typicality values using;

$$t_{ij} = \frac{1}{1 + \left(\frac{\left\|x_j - v_i\right\|^2}{Y_i}\right)^{\frac{1}{m-1}}}$$

End the loop for typicality values;
In next step put the terminating condition for checking optimization of solution.
End the "For" loop for number of iterations 'r';
In next step find the value of objective function;

$$J_{PCM} = \sum_{j=1}^{N}\sum_{i=1}^{C}(t_{ij})^m \|x_j - v_i\|^2 + \sum_{i=1}^{C} Y_i \sum_{j=1}^{N}(1 - t_{ij})^m$$

And plot the data points and centers on the graph for better analysis of data.

3.3 Python code

```
python
CopyEdit
import numpy as np
import matplotlib.pyplot as plt

# Set parameters
m = 2 # fuzzifier
C = 3 # number of clusters
N = 100 # number of data points
K = 1 # constant for Y_i calculation
r = 50 # number of iterations

# Generate random data
np.random.seed(0)
X = np.vstack((np.random.randn(N//2, 2) + [2, 2], np.random.randn(N//2,
2) + [-2, -2]))

# Initialize typicality matrix randomly
t = np.random.rand(C, N)

# Main PCM loop
for iteration in range(r):
    # Compute cluster centers
    v = np.zeros((C, X.shape[1]))
```

```
    for i in range(C):
        numerator = np.sum((t[i] ** m).reshape(-1, 1) * X, axis=0)
        denominator = np.sum(t[i] ** m)
        v[i] = numerator / denominator

    # Compute distances
    d = np.zeros((C, N))
    for i in range(C):
        d[i] = np.linalg.norm(X - v[i], axis=1)

    # Compute Y_i
    Y = np.zeros(C)
    for i in range(C):
        num = K * np.sum((t[i] ** m) * (d[i] ** 2))
        denom = np.sum(t[i] ** m)
        Y[i] = num / denom

    # Update typicality values
    for i in range(C):
        t[i] = 1 / (1 + (d[i] ** 2 / Y[i]) ** (1 / (m - 1)))

# Compute objective function
J_PCM = 0
for i in range(C):
    J_PCM += np.sum((t[i] ** m) * (d[i] ** 2)) + Y[i] * np.sum((1 -
t[i]) ** m)

print("Final objective function value:", J_PCM)

# Plot data and cluster centers
plt.scatter(X[:, 0], X[:, 1], c='grey', alpha=0.5, label='Data points')
plt.scatter(v[:, 0], v[:, 1], c='red', marker='X', s=200, label='Cluster
centers')
plt.legend()
plt.title('PCM Clustering Result')
plt.show()
```

3.4 R code

```
set.seed(0)
library(ggplot2)

# Parameters
m <- 2
C <- 3
N <- 100
K <- 1
r <- 50

# Generate random data
X <- rbind(matrix(rnorm(N, mean=2), ncol=2),
       matrix(rnorm(N, mean=-2), ncol=2))

# Initialize typicality matrix
t <- matrix(runif(C * nrow(X)), nrow=C)

# Main loop
for (iter in 1:r) {
 # Compute cluster centers
 v <- matrix(0, nrow=C, ncol=ncol(X))
 for (i in 1:C) {
  numerator <- colSums((t[i,]^m) * X)
  denominator <- sum(t[i,]^m)
  v[i, ] <- numerator / denominator
 }

 # Compute distances
 d <- matrix(0, nrow=C, ncol=nrow(X))
 for (i in 1:C) {
  d[i, ] <- sqrt(rowSums((X - matrix(v[i, ], nrow=nrow(X), ncol=
  ncol(X), byrow=TRUE))^2))
 }

 # Compute Y_i
 Y <- numeric(C)
 for (i in 1:C) {
  num <- K * sum((t[i,]^m) * (d[i,]^2))
```

```
    denom <- sum(t[i,]^m)
    Y[i] <- num / denom
   }

   # Update typicality
   for (i in 1:C) {
    t[i, ] <- 1 / (1 + (d[i,]^2 / Y[i])^(1 / (m - 1)))
   }
  }

  # Compute objective function
  J_PCM <- 0
  for (i in 1:C) {
   J_PCM <- J_PCM + sum((t[i,]^m) * (d[i,]^2)) + Y[i] * sum((1 -
   t[i,])^m)
  }

  print(paste("Final objective function value:", J_PCM))

  # Plot
  df <- data.frame(X1=X[,1], X2=X[,2])
  centers <- data.frame(X1=v[,1], X2=v[,2])

  ggplot(df, aes(x=X1, y=X2)) +
   geom_point(alpha=0.5) +
   geom_point(data=centers, color='red', size=4, shape=4) +
ggtitle("PCM Clustering Result")
```

3.5 Solved examples

Example 1

The study investigates 15 CSK IPL players' stats including batting averages, over rates, and selection positions shown in Table 3.1. Employing the PCM clustering technique, we want to categorize the players into three unique clusters. Our analysis groups players into batters, all-rounders, and fast bowlers. Through this technique, we discover the strongest bowling all-rounder performers within our dataset.

Table 3.1: CSK team members' batting and bowling statistics along with their early designations.

Player	Batting average	Bowl average	Initial class
MM Ali	24.96	22.78	Batsman
S Anirudha	19	50	Bowler
R Ashwin	9.5	24.22	Bowler
S Badrinath	30.65	50	Batsman
L Balaji	3.66	26.09	Bowler
DE Bollinger	21	18.72	Batsman
DJ Bravo	25.1	22.47	Batsman
DL Chahar	10.83	28.2	Bowler
SM Curran	22	30.54	Batsman
MS Dhoni	40.53	50	Batsman
F du Plessis	35.33	50	Batsman
RD Gaikwad	39.36	50	Batsman
MS Gony	12.5	30.73	Bowler
Harbhajan Singh	7.5	25.3	Bowler

Research inquiries:

(i) **Classification of players:**
What is the use of PCM clustering for categorizing players into three distinct categories based on their batting and bowling averages: Our system identifies players into three separate groups, which are batsmen, bowling all-rounders, and bowlers.

(ii) **Determination of optimal bowling all-rounders:**
What are the profiles of players in our sample who perform well both as bowlers and batsmen, and how do they stack up versus other player types?

(iii) **Cluster analysis:**
How does each performance segment appear different through its bowling and batting statistics?

The discovered clusters show how much they match the player categories from step one.

(iv) **Performance analysis:**
By using PCM clustering, we can find player performance patterns that we cannot see through simple data review.

Solution:

Step 1: Consider batting average = X1 and bowling average = X2.

X1	24.96	19	9.5	30.65	3.66	21	25.1	10.83	22	40.53	35.33	39.36	12.5	7.5
X2	22.78	50	24.22	50	26.09	18.72	22.47	28.2	30.54	50	50	50	30.73	25.3

Step 2: Starting partition of the membership function into three clustering centres.

Mu1	0.009	0.233	0.954	0.019	0.914	0.113	0.012	0.972	0.213	0.019	0.000	0.012	0.851	0.976
Mu2	0.003	0.566	0.006	0.957	0.017	0.020	0.004	0.004	0.070	0.953	0.996	0.969	0.025	0.004
Mu3	0.988	0.200	0.040	0.024	0.068	0.867	0.984	0.024	0.717	0.029	0.000	0.019	0.125	0.020

Step 3: Take $m = 2$.

Mu1^2	Mu2^2	Mu3^2	X1*Mu1^2	X2*Mu1^2	X1*Mu2^2	X2*Mu2^2	X1*Mu3^2	X2*Mu3^2
8.91E-05	9.32E-06	0.975172	0.002224	0.002029637	0.000233	0.000212	24.34029	22.21441982
0.05444	0.320807	0.040111	1.034363	2.722008372	6.095334	16.04035	0.762111	2.005554055
0.909592	3.58E-05	0.001623	8.641121	22.03031173	0.00034	0.000868	0.01542	0.039312236
0.00035	0.91563	0.000596	0.010713	0.01747694	28.06405	45.78148	0.018276	0.029813494
0.83604	0.000306	0.004645	3.059906	21.8122833	0.00112	0.007985	0.017	0.121184326
0.012696	0.000417	0.75152	0.266622	0.237674597	0.008756	0.007806	15.78191	14.06844884
0.000152	1.62E-05	0.96755	0.003817	0.003417342	0.000407	0.000364	24.28549	21.74083743
0.944111	1.74E-05	0.000585	10.22472	26.62392766	0.000188	0.00049	0.006331	0.01648642
0.045331	0.004912	0.514096	0.997284	1.384412102	0.108059	0.150005	11.31012	15.70050067
0.000353	0.907604	0.000814	0.014316	0.017660613	36.78521	45.38022	0.032974	0.040678988
0	0.991394	0	0	0	35.02595	49.56971	0	0
0.000154	0.938844	0.000348	0.006059	0.00769727	36.95291	46.94221	0.013694	0.017396339
0.723552	0.000611	0.01554	9.044404	22.23476191	0.007638	0.018778	0.194256	0.47755799
0.951905	1.5E-05	0.000419	7.139289	24.08320183	0.000112	0.000379	0.003144	0.010607031
4.478766	**4.080619**	**3.273019**	**40.44484**	**121.1768633**	**143.0503**	**203.9009**	**76.78103**	**76.48279764**

Step 4: Value of cluster centers

v1	07.50000	25.30000
v2	35.33000	50.0000
v3	24.9600	22.7800

Step 5: Distances of data points from all cluster centers

D1	311.202	742.34	5.1663	1146.01	15.369	225.54	317.768	19.4989	237.707	1701.070	1384.598	1625.149	54.4849	6.98e-18
D2	848.46	266.66	1331.79	21.902	1574.6	1183.78	862.553	1075.48	556.380	27.040	3.74e-17	16.2409	892.54	1384.59
D3	1.141e-17	776.45	241.08	773.30	464.64	32.165	0.11569	229.033	68.979	983.353	848.46	948.288	218.45	311.20

Step 6: Final membership values

NMu1	1.0e-10	1.0e-10	1.0e-10	1.0e-10	1.0e-10	1.0e-10	1.0e-10	1.0e-10	1.0e-10	1.0e-10	1.0e-10	1.0e-10	1.0e-10	0.999999973312575
NMu2	1.0e-10	1.0e-10	1.0e-10	1.0e-10	1.0e-10	1.0e-10	1.0e-10	1.0e-10	1.0e-10	1.0e-10	1	1.0e-10	1.0e-10	1.0e-10
NMu3	1	1.0e-10	1.0e-10	1.0e-10	1.0e-10	1.0e-10	8.643e-10	1.0e-10	1.0e-10	1.0e-10	1.0e-10	1.0e-10	1.0e-10	0.999999973312575

Step 7: $e = 0.0001$.

e1	1.2346e-05	1.2902e-05	1.0898e-05	2.4337e-06	1.4862e-05	1.2242e-05	1.0961e-05	1.4304e-05	1.3675e-05	2.4236e-06	3.0403e-09	1.0961e-05	1.4724e-05	1.4305e-05
e2	1.2725e-05	1.1845e-05	1.2697e-05	1.3711e-05	1.2567e-05	1.2396e-05	1.2758e-05	1.2759e-05	1.2964e-05	1.3706e-05	1.3774e-05	1.3746e-05	1.2935e-05	1.2759e-05
e3	3.8287e-06	1.2903e-05	1.2697e-05	1.3705e-05	1.2568e-05	3.7152e-06	3.8749e-06	1.3706e-05	1.3326e–05	1.3706e-05	3.0403e-09	1.0961e-05	1.2935e-05	1.2759e-05

Example 2

An organization has warehouses for life-saving drugs at six different locations, whose coordinates are given in the following table. The organization is in the process of locating one more critical central warehouse, which will distribute drugs to all the existing warehouses on emergency requests. Find the number of optimum locations of the new facility (warehouse) based on PCM clustering concept.

Table 3.2: Thirty warehouses whose coordinates are summarized.

Existing warehouse number	Coordinates of centroids
1	(2.5,3.5)
2	(2.7,3.6)
3	(2.2,3.2)
4	(3.0,3.0)
5	(2.7,3.1)
6	(2.3,3.5)
7	(3.1,3.0)
8	(3.3,2.9)
9	(3.0,3.5)
10	(2.9, 2.9)
11	(6.0,7.2)
12	(6.5,7.0)
13	(5.9,6.9)
14	(6.3,7.3)
15	(5.9,7.0)
16	(6.8,7.0)
17	(6.5,6.9)
18	(6.2,7.1)
19	(7.2,7.2)
20	(6.6,7.1)
21	(11.1,13.5)
22	(12.2,12.5)
23	(11.5,11.9)

Solution:

The data are shown in Table 3.2. Assume that we want to determine a fuzzy pseudo-partition with three clusters (i.e. $c = 3$). Assume further that we choose $m = 2$. $||\,.\,||$ is the Euclidean distance, and the initial fuzzy pseudo-partition is $U^{(0)} = [U_1, U_2, U_3]$. Table 3.3–3.7 shows the respective results obtained at different iterations.

Table 3.3: Cluster centres and the corresponding membership function values of matrix U.

Input data		Membership function for fuzzy clustering		
X_1	X_2	U_1	U_2	U_3
2.5	3.5	.9	.05	.05
2.7	3.6	.89	.055	.055
2.2	3.2	.88	.06	.06
3.0	3.0	.87	.065	.065
2.7	3.1	.86	.07	.07
2.3	3.5	.84	.09	.07
3.1	3.0	.81	.11	.08
3.3	2.9	.8	.1	.1
3.0	3.5	.79	.11	.1
2.9	2.9	.75	.15	.1
6.0	7.2	.05	.9	.05
6.5	7.0	.055	.89	.055
5.9	6.9	.06	.88	.06
6.3	7.3	.065	.87	.065
5.9	7.0	.07	.86	.07
6.8	7.0	.09	.84	.07
6.5	6.9	.11	.81	.08
6.2	7.1	.1	.8	.1
7.2	7.2	.11	.79	.1
6.6	7.1	.15	.75	.1
11.1	13.5	.05	.05	.9
12.2	12.5	.055	.055	.89
11.5	11.9	.06	.06	.88
13.2	13.3	.065	.065	.87
11.5	12.5	.07	.07	.86
12.0	15.0	.07	.09	.84
13.2	13.2	.08	.11	.81
11.2	11.9	.1	.1	.8
12.9	13.7	.1	.11	.79
13.9	13.0	.1	.15	.75

Iteration 1:

Table 3.4: Cluster centres V and the corresponding membership function values of matrix U (Iteration 1).

Cluster centres		V_1	V_2	V_3
		(4.2002, 4.3096)	**(8.0911, 8.1712)**	**(10.8775, 10.9743)**
Average values		**(4.20, 4.31)**	**(8.09, 8.17)**	**(10.88, 10.97)**
Data points		U_1	U_2	U_3
2.5	3.5	.9973	.0026	.0001
2.7	3.6	.9999	.0000	.0000
2.2	3.2	.9999	.0000	.0000
3.0	3.0	.9995	.0004	.0001
2.7	3.1	.9999	.0000	.0000
2.3	3.5	.9969	.0028	.0002
3.1	3.0	.9952	.0046	.0002
3.3	2.9	.9997	.0002	.0001
3.0	3.5	.9999	.0000	.0000
2.9	2.9	.9998	.0001	.0001
6.0	7.2	.0001	.9973	.0025
6.5	7.0	.0001	.9980	.0018
5.9	6.9	.0001	.9991	.0007
6.3	7.3	.0002	.9970	.0027
5.9	7.0	.0001	.9991	.0007
6.8	7.0	.0004	.9955	.0041
6.5	6.9	.0003	.9963	.0033
6.2	7.1	.0003	.9970	.0027
7.2	7.2	.0003	.9995	.0002
6.6	7.1	.0002	.9995	.0002
11.1	13.5	.0000	.0001	.9998
12.2	12.5	.0000	.0001	.9998
11.5	11.9	.0000	.0002	.9997
13.2	13.3	.0000	.0001	.9998
11.5	12.5	.0000	.0002	.9997
12.0	15.0	.0001	.0002	.9997
13.2	13.2	.0002	.0008	.9990
11.2	11.9	.0004	.0022	.9974
12.9	13.7	.0002	.0002	.9996
13.9	13.0	.0002	.0002	.9996

Iteration 2:

Table 3.5: Cluster centres V and the corresponding membership function values of matrix U (Iteration 2).

Cluster centres		V_1	V_2	V_3
		(2.8015, 3.160)	**(6.3746, 7.082)**	**(12.0207, 12.768)**
Average values		**(2.80, 3.16)**	**(6.37, 7.08)**	**(12.02, 12.77)**
Data points		U_1	U_2	U_3
2.5	3.5	.9461	.0337	.0242
2.7	3.6	.9358	.0388	.0254
2.2	3.2	.9691	.0186	.0122
3.0	3.0	.8712	.0896	.0392
2.7	3.1	.9258	.0492	.0250
2.3	3.5	.9612	.0252	.0136
3.1	3.0	.8791	.0853	.0356
3.3	2.9	.8281	.1242	.0477
3.0	3.5	.8747	.0901	.0352
2.9	2.9	.8974	.0743	.0283
6.0	7.2	.0255	.9373	.0372
6.5	7.0	.0173	.9516	.0311
5.9	6.9	.0313	.9401	.0286
6.3	7.3	.0190	.9524	.0286
5.9	7.0	.0320	.9388	.0292
6.8	7.0	.0110	.9584	.0306
6.5	6.9	.0171	.9533	.0296
6.2	7.1	.0243	.9439	.0318
7.2	7.2	.0052	.9658	.0290
6.6	7.1	.0152	.9516	.0332
11.1	13.5	.0050	.0211	.9739
12.2	12.5	.0043	.0262	.9695
11.5	11.9	.0054	.0330	.9616
13.2	13.3	.0026	.0197	.9777
11.5	12.5	.0054	.0329	.9617
12.0	15.0	.0031	.0176	.9793
13.2	13.2	.0024	.0192	.9784
11.2	11.9	.0053	.0340	.9607
12.9	13.7	.0023	.0173	.9804
13.9	13.0	.0015	.0147	.9838

Due to space and readability considerations, it is impractical to present the complete set of iterative results. Therefore, the following tables provide a summary of the membership values and cluster centres obtained during the penultimate (i.e. 6th) and final (i.e. 7th) iterations, which effectively illustrate the convergence behaviour of the algorithm.

Iteration 6:

Table 3.6: Cluster centres V and the corresponding membership function values of matrix U (Iteration 6).

Cluster centres		V_1	V_2	V_3
		(5.57, 6.92)	**(11.52, 12.86)**	**(12.89, 12.58)**
Average values		**(5.57, 6.92)**	**(11.52, 12.86)**	**(12.89, 12.58)**
Data points		U_1	U_2	U_3
2.5	3.5	.8999	.0499	.0502
2.7	3.6	.8897	.0552	.0551
2.2	3.2	.8804	.0600	.0596
3.0	3.0	.8699	.0651	.0648
2.7	3.1	.8598	.0703	.0699
2.3	3.5	.8397	.0905	.0698
3.1	3.0	.8110	.1101	.0790
3.3	2.9	.7995	.1009	.0996
3.0	3.5	.7903	.1111	.0986
2.9	2.9	.7499	.1500	.1001
6.0	7.2	.0501	.8999	.0500
6.5	7.0	.0553	.8898	.0550
5.9	6.9	.0598	.8804	.0598
6.3	7.3	.0647	.8697	.0656
5.9	7.0	.0698	.8602	.0699
6.8	7.0	.0899	.0700	.8401
6.5	6.9	.1102	.0798	.8100
6.2	7.1	.0999	.1009	.7992
7.2	7.2	.1098	.0988	.7914
6.6	7.1	.1502	.1011	.7497
11.1	13.5	.0500	.0503	.8998
12.2	12.5	.0549	.0551	.8899
11.5	11.9	.0600	.0598	.8802
13.2	13.3	.0654	.0645	.8698
11.5	12.5	.0697	.0700	.8603
12.0	15.0	.0699	.0701	.8398
13.2	13.2	.0799	.0789	.8112
11.2	11.9	.0997	.1007	.7996
12.9	13.7	.0989	.0988	.7923
13.9	13.0	.0998	.1002	.7998

Iteration 7:

Table 3.7: Cluster centres *V* and the corresponding membership function values of matrix *U* (Iteration 7).

Cluster centres		V_1	V_2	V_3
		(5.57, 6.92)	(11.52, 12.86)	(12.89, 12.58)
Average values		(5.57, 6.92)	(11.52, 12.86)	(12.89, 12.58)
Data points		U_1	U_2	U_3
2.5	3.5	.8997	.0501	.0502
2.7	3.6	.8899	.0551	.0550
2.2	3.2	.8803	.0602	.0595
3.0	3.0	.8697	.0655	.0648
2.7	3.1	.8600	.0701	.0699
2.3	3.5	.8398	.0904	.0698
3.1	3.0	.8109	.1100	.0791
3.3	2.9	.7990	.0999	.1011
3.0	3.5	.7904	.1108	.0988
2.9	2.9	.7500	.1501	.0999
6.0	7.2	.0500	.8998	.0502
6.5	7.0	.0550	.8899	.0551
5.9	6.9	.0602	.8804	.0594
6.3	7.3	.0650	.8699	.0651
5.9	7.0	.0700	.8601	.0699
6.8	7.0	.0900	.0698	.8402
6.5	6.9	.1100	.0809	.8091
6.2	7.1	.1000	.1010	.7990
7.2	7.2	.1100	.0989	.7901
6.6	7.1	.1500	.1002	.7498
11.1	13.5	.0500	.0502	.8998
12.2	12.5	.0550	.0550	.8899
11.5	11.9	.0601	.0595	.8803
13.2	13.3	.0650	.0648	.8698
11.5	12.5	.0700	.0699	.8601
12.0	15.0	.0700	.0698	.8402
13.2	13.2	.0800	.0791	.8109
11.2	11.9	.1000	.1011	.7989
12.9	13.7	.1000	.0988	.7912
13.9	13.0	.1000	.0998	.7999

Then, the algorithm stops after seven iterations, because max $\left\{\left|u_{ij}^{k+1} - u_{ij}^{k}\right|\right\} < 8.28 \times 10^{-5}$ and we obtain the pseudo-partition defined in Table 2.7. The three clusters centres are $V_1 = (5.57, 6.92)$, $V_2 = (11.52, 12.86)$, and $V_3 = (12.89, 12.58)$, and $J_{PCM} = 3.0954$.

3.6 Practice set

Problem 1

The study evaluates five football players based on their **passing accuracy** and **tackles won**. Employing the **PCM** clustering technique, we aim to categorize the players into **three role-based clusters**. This analysis groups players into **playmakers, defenders**, and **all-rounders**. Through this technique, we also investigate the **typicality values**, which represent how strongly a player fits into a given cluster – helping us handle uncertain or noisy data better.

Table 1: Football players' passing accuracy and tackles won data.

Player	Passing accuracy (%)	Tackles won (%)
Player A	88.5	32.0
Player B	72.0	85.5
Player C	80.0	70.0
Player D	91.2	29.8
Player E	60.0	88.0

Research inquiries:

(i) **Role classification using PCM:**
How does the PCM clustering technique classify players into three different groups (playmakers, defenders, and all-rounders) based on their passing and tackling statistics?

(ii) **Handling uncertainty and noise:**
Which players demonstrate high **typicality** to a cluster, and which ones are borderline? How does PCM deal with outlier or noisy players compared to FCM?

(iii) **Player profiling:**
Based on the cluster centres obtained, identify the defining characteristics of each cluster. Who are the **pure** players (high typicality), and who belong partially to more than one role?

(iv) **Insight beyond thresholding:**
What kind of hidden or overlapping player roles can PCM help discover that would be difficult to identify using traditional cut-off-based methods?

Problem 2

This study analyses the **academic profiles** of five students using three performance indicators: **Math score, science score**, and **English score**. Using the **PCM** clustering technique, we aim to categorize students into **three academic profiles**:

- **STEM-focused learners** (strong in maths and science)
- **Language-focused learners** (strong in English)
- **All-rounders** (balanced in all subjects)

PCM enables the identification of **typicality values** for each student in the clusters, helping detect students who don't clearly belong to a single category.

Table 2: Student scores across three subjects.

Student	Math score	Science score	English score
Student_1	90	92	65
Student_2	60	55	88
Student_3	78	75	70
Student_4	92	94	68
Student_5	58	50	90

Research inquiries:

(i) **Profile-based grouping:**
How does the PCM algorithm group students into **STEM-focused**, **language-focused**, and **all-rounder** categories based on their three subject scores?

(ii) **Typicality analysis:**
Which students show **high typicality** to a particular profile, and which are **borderline** or **uncertain**? How does PCM handle mixed or overlapping student behaviours?

(iii) **Cluster centre interpretation:**
What do the cluster centres reveal about the academic focus of each group? How different are the performance patterns of each cluster?

(iv) **Application of PCM in education:**
How can educational institutions use PCM clustering to identify learning strengths in students more flexibly than traditional averaging or grading methods?

Problem 3

This study assesses the **in-game performance** of five e-sports players based on four key metrics: **Kills, assists, deaths,** and **damage taken** during matches. Using the **PCM** clustering algorithm, players are grouped into the following categories:

- **Attackers** (high kills and damage dealt)
- **Supporters** (high assists, low kills)
- **Tanks** (high damage taken and survival focus)

PCM helps us understand **typicality levels** – how well a player fits a role – and better handles noise or uncertain performance overlaps.

Table 3: E-sports player in-game statistics.

Player	Kills	Assists	Deaths	Damage taken
Player A	18	4	3	2,100
Player B	4	17	2	1,100
Player C	12	8	6	3,000
Player D	3	15	1	900
Player E	6	6	9	4,800

Research inquiries:

(i) **Cluster-based role detection:**
How does the PCM algorithm classify players into **attackers**, **supporters**, and **tanks** using the four-performance metrics?

(ii) **Typicality interpretation:**
Which players are **strongly assigned** to a cluster, and which have **low typicality** indicating hybrid or unclear roles?

(iii) **Cluster profile comparison:**
By analysing the cluster centres, what can we infer about the statistical signature of each player role in e-sports?

(iv) **PCM advantage over hard labels:**
How does PCM provide a more realistic and flexible understanding of player roles compared to hard categorization based on thresholds?

Chapter 4
Robust fuzzy *C*-means (FCM-σ)

4.1 Introduction

Robust fuzzy *C*-means (FCM-σ) is a modification of the standard fuzzy *C*-means (FCM) clustering algorithm, which is specifically meant to clustering noisy data and outliers better. Although standard FCM calculates the membership values only depending on the distances of data points with cluster centres, it is sensitive to points that are distant to the cluster centres which may corrupt the output of the clustering outcome.

To deal with it, FCM-σ presents a cluster-wise scale parameter sigma; that is, the spread (or standard deviation) of the data in each cluster. FCM-σ alleviates the role of far-away or contaminated points in the update calculation of the cluster centres because the contributions of the far-away points and noisy points are minimized by involving them in the computation of the membership value and objective functions. Such self-adaptive weighting stabilizes and improves the reliability of the clustering process and ensures that clusters reflect the actual data structure all the more, even when there are outliers or clusters with different densities.

This technique is introduced as a means to enhance the performance of the FCM algorithm while dealing with datasets that exhibit clusters of inconsistent densities and non-hyper spherical forms [29, 37, 43]. The objective function of this method is the inclusion of a normalized distance component, which is defined as follows:

$$J_{\mathrm{FCM}-\sigma} = \sum_{j=1}^{N}\sum_{i=1}^{C}\mu_{ij}^{m}\,\frac{\left\|x_j - v_i\right\|^2}{\sigma_i{}^2};\ \sum_{i=1}^{C}\mu_{ij} = 1$$

where

$$\sigma_i{}^2 = \frac{\sum_{j=1}^{N}\mu_{ij}^{m}\left\|x_j - v_i\right\|^2}{\sum_{j=1}^{N}\mu_{ij}^{m}}$$

The algorithm employed in this study utilizes a normalization technique to account for the distribution of data from each cluster centre. Cluster centre v_i for the function to be optimized is the same as that of FCM [44–46]. Thus, the updated cluster centre and membership grade are as follows:

$$v_i = \frac{\sum_{j=1}^{N}\mu_{ij}^{m}x_j}{\sum_{j=1}^{N}\mu_{ij}^{m}}$$

https://doi.org/10.1515/9783111494272-004

$$\mu_{ij} = \left[\sum_{k=1}^{C} \left(\frac{\frac{||x_j - v_i^2||}{\sigma_i^2}}{\frac{x_j - v_k^2}{\sigma_k^2}} \right)^{\frac{1}{m-1}} \right]^{-1}$$

This algorithm commences by initializing the matrix $U_{N \times C}$ with random values. Subsequently, in each iteration σ_i, the values of μ_{ij} and v_i are adjusted iteratively until convergence is achieved.

4.2 Mathematical derivation for robust FCM

Derivation for membership μ_{ik}

For finding membership, we will use Lagrangian multiplier method. The Lagrangian equation for FCM-σ is

$$L = \sum_{j=1}^{N} \left(\sum_{i=1}^{C} \mu_{ij}^{m} \frac{||x_j - v_i||^2}{\sigma_i^2} - \lambda \left(\sum_{i=1}^{C} \mu_{ij} - 1 \right) \right)$$

Take partial derivative w.r.t μ_{ij}:

$$\frac{\partial L}{\partial \mu_{ij}} = m\mu_{ij}^{m-1} \frac{||x_j - v_i||^2}{\sigma_i^2} - \lambda$$

Set it to zero:

$$=> m\,\mu_{ij}^{m-1} \frac{||x_j - v_i||^2}{\sigma_i^2} - \lambda = 0$$

$$=> m\,\mu_{ij}^{m-1} \frac{||x_j - v_i||^2}{\sigma_i^2} = \lambda$$

$$=> \mu_{ij}^{m-1} = \frac{\lambda \sigma_i^2}{m\,||x_j - v_i||^2}$$

$$=> \mu_{ij} = \left(\frac{\lambda \sigma_i^2}{m\,||x_j - v_i||^2} \right)^{\frac{1}{m-1}} \tag{4.1}$$

To eliminate λ, we apply the normalization constraint:

$$\sum_{i=1}^{C} \mu_{ij} = 1 \tag{4.2}$$

So, from (4.1) and (4.2):

$$\sum_{i=1}^{C} \mu_{ij} = \sum_{i=1}^{C} \left(\frac{\lambda \sigma_i^2}{m \, \|x_j - v_i\|^2} \right)^{\frac{1}{m-1}} = 1$$

$$=> \left(\frac{\lambda}{m} \right)^{\frac{1}{m-1}} = \left[\sum_{i=1}^{C} \left(\frac{\sigma_i^2}{\|x_j - v_i\|^2} \right)^{\frac{1}{m-1}} \right]^{-1}$$

Putting this value in (4.1), we get

$$\mu_{ij} = \left(\frac{\sigma_i^2}{\|x_j - v_i\|^2} \right)^{\frac{1}{m-1}} \left[\sum_{k=1}^{C} \left(\frac{\sigma_k^2}{\|x_j - v_k\|^2} \right)^{\frac{1}{m-1}} \right]^{-1}$$

$$\mu_{ij} = \left[\sum_{k=1}^{C} \left(\frac{\frac{\|x_j - v_i\|^2}{\sigma_i^2}}{\frac{\|x_j - v_k\|^2}{\sigma_k^2}} \right)^{\frac{1}{m-1}} \right]^{-1}$$

Derivation for cluster centre v_i

For the cluster centre, we compute the partial derivative of $J_{\text{FCM}-\sigma}$ w.r.t v_i:

$$\frac{\partial J_{\text{FCM}-\sigma}}{\partial v_i} = \sum_{j=1}^{N} -2 \, \mu_{ij}^m \frac{(x_j - v_i)}{\sigma_i^2}$$

Set it to zero:

$$\frac{-2}{\sigma_i^2} \sum_{j=1}^{N} \mu_{ij}^m (x_j - v_i) = 0$$

$$=> \sum_{j=1}^{N} \mu_{ij}^m x_j = v_i \sum_{j=1}^{N} \mu_{ij}^m$$

$$=> v_i = \frac{\sum_{j=1}^{N} \mu_{ij}^m x_j}{\sum_{j=1}^{N} \mu_{ij}^m}$$

Derivation of σ_i^2

For finding σ_i^2 partial derivative would not help. So, instead of relying on setting the derivative to zero, we can recognize that the objective function can be seen as weighted sum of squared distances. So, to minimize $J_{FCM-\sigma}$, the best of value of σ_i^2 will be the one that balances this weighted sum.

The optimal value of σ_i^2 can be obtained by recognizing that the scale σ_i^2 corresponding to a weighted average of the square distance. Thus, we can directly write

$$\sigma_i^2 = \frac{\sum_{j=1}^{N} \mu_{ij}^m \,||x_j - v_i||^2}{\sum_{j=1}^{N} \mu_{ij}^m}$$

FCM-σ:

---

m=2;

Choose number of cluster 'C' center convenient for your data;

Initially take random C×N membership value matrix;

Start "For" loop for number of iterations (say r);

Take data set;

Find initial cluster centers using formula;

$$v_i = \frac{\sum_{j=1}^{N} \mu_{ij}^m x_j}{\sum_{j=1}^{N} \mu_{ij}^m}$$

Find distance 'd' between cluster centers and data points;

Find value of ' σ ' using;

$$\sigma_i = \left(\frac{\sum_{j=1}^{N} \mu_{ij}^m \,||x_j - v_i||^2}{\sum_{j=1}^{N} \mu_{ij}^m}\right)^{\frac{1}{2}}$$

Again start "For" loop for finding membership values iteratively as many times as number of data points.

Find membership values using;

$$\mu_{ij} = \left[\sum_{k=1}^{C}\left(\frac{\frac{||x_j - v_i||^2}{\sigma_i}}{\frac{||x_j - v_k||^2}{\sigma_k}}\right)^{\frac{1}{m-1}}\right]^{-1}$$

End the loop for membership values;

In next step put the terminating condition for checking optimization of solution.

End the "For" loop for number of iterations 'r';

In next step find the value of objective function;

$$J_{FCM-\sigma} = \sum_{j=1}^{N}\sum_{i=1}^{C} \mu_{ij}^m \frac{||x_j - v_i||^2}{\sigma_i}$$

And plot the data points and centers on the graph for better analysis of data.

4.3 Python Code

```
import numpy as np
import matplotlib.pyplot as plt
# Parameters
m = 2
C = 3   # number of clusters
r = 50  # number of iterations
N = 150 # number of data points

# Generate synthetic data
np.random.seed(0)
X = np.vstack([
np.random.randn(N//3, 2) + [2, 2],
np.random.randn(N//3, 2) + [-2, -2],
np.random.randn(N//3, 2) + [4, -3]
])

# Initialize random membership matrix
mu = np.random.rand(C, len(X))
mu = mu / np.sum(mu, axis=0, keepdims=True) # normalize over clusters

for iteration in range(r):
# Compute cluster centers
v = np.zeros((C, X.shape[1]))
for i in range(C):
numerator = np.sum((mu[i]**m).reshape(-1,1) * X, axis=0)
denominator = np.sum(mu[i]**m)
v[i] = numerator / denominator

# Compute distances
d = np.zeros((C, len(X)))
for i in range(C):
d[i] = np.linalg.norm(X - v[i], axis=1)

# Compute sigma_i for each cluster
sigma = np.zeros(C)
for i in range(C):
numerator = np.sum((mu[i]**m) * (d[i]**2))
denominator = np.sum(mu[i]**m)
sigma[i] = np.sqrt(numerator / denominator + 1e-8) # add small value to
avoid zero division
```

```
# Update membership values
for j in range(len(X)):
for i in range(C):
denom = 0
for k in range(C):
ratio = (d[i,j]**2 / sigma[i]) / (d[k,j]**2 / sigma[k] + 1e-8)
denom += ratio**(1/(m-1))
mu[i,j] = 1 / (denom + 1e-8)

# Compute objective function
J = 0
for i in range(C):
J += np.sum((mu[i]**m) * (d[i]**2) / sigma[i])

print("Final objective function value:", J)

# Plot result
plt.scatter(X[:,0], X[:,1], c=np.argmax(mu, axis=0), cmap='viridis',
alpha=0.5)
plt.scatter(v[:,0], v[:,1], c='red', s=200, marker='X', label='Centers')
plt.title('Modified FCM with σ')
plt.legend()
plt.show()
```

4.4 R code

```
set.seed(42)
modified_fcm_sigma <- function(X, C=3, m=2, max_iter=100, tol=1e-5) {
 N <- nrow(X)
 D <- ncol(X)
 mu <- matrix(runif(C * N), nrow=C)
 mu <- mu / colSums(mu)

 obj_values <- numeric(max_iter)

 for (iter in 1:max_iter) {
  # Step 1: compute centers
  v <- matrix(0, nrow=C, ncol=D)
  for (i in 1:C) {
   num <- colSums((mu[i,]^m) * X)
   denom <- sum(mu[i,]^m)
```

```
 v[i,] <- num / (denom + 1e-8)
}

# Step 2: compute distances
d <- matrix(0, nrow=C, ncol=N)
for (i in 1:C) {
 d[i,] <- sqrt(rowSums((X - matrix(v[i,], nrow=N, ncol=D,
 byrow=TRUE))^2))
}

# Step 3: compute sigma_i
sigma <- numeric(C)
for (i in 1:C) {
 numerator <- sum((mu[i,]^m) * d[i,]^2)
 denominator <- sum(mu[i,]^m)
 sigma[i] <- sqrt(numerator / (denominator + 1e-8))
}

# Step 4: update membership
mu_new <- matrix(0, nrow=C, ncol=N)
for (j in 1:N) {
 for (i in 1:C) {
  denom <- 0
  for (k in 1:C) {
   ratio <- (d[i,j]^2 / sigma[i]) / (d[k,j]^2 / sigma[k] + 1e-8)
   denom <- denom + ratio^(1/(m-1))
  }
  mu_new[i,j] <- 1 / (denom + 1e-8)
 }
}

# Step 5: compute objective function
J <- sum(sapply(1:C, function(i) sum((mu_new[i,]^m) * d[i,]^2 /
(sigma[i]+1e-8))))
obj_values[iter] <- J

if (sqrt(sum((mu_new - mu)^2)) < tol) {
 cat("Converged at iteration", iter, "\n")
 break
}
```

```
  mu <- mu_new
 }

 # Plot objective
 plot(obj_values[1:iter], type="o", main="Objective Function over
 Iterations",
   xlab="Iteration", ylab="Objective")

 # Plot clustering result
 cluster <- apply(mu, 2, which.max)
 plot(X, col=cluster, pch=19, main="Clustering Result")
 points(v, col=1:C, pch=8, cex=2)

 list(centers=v, membership=mu, sigma=sigma, objective=obj_values
 [1:iter])
}

# Example usage:
X <- rbind(
 matrix(rnorm(100, mean=2), ncol=2),
 matrix(rnorm(100, mean=-2), ncol=2),
 matrix(rnorm(100, mean=4), ncol=2)
)

result <- modified_fcm_sigma(X, C=3, max_iter=50)
print(result$centers)
print(result$sigma)
```

4.5 Solved examples

Example 1

The study investigates 15 CSK IPL players' stats, including batting averages, over rates, and selection positions shown in Table 4.1. Employing the FCM-σ clustering technique, we want to categorize the players into three unique clusters. Our analysis groups players into batters, all-rounders, and fast bowlers. Through this technique, we discover the strongest bowling all-rounder performers within our dataset.

Table 4.1: CSK team members' batting and bowling statistics along with their early designations.

Player	Batting average	Bowl average	Initial class
MM Ali	24.96	22.78	Batsman
S Anirudha	19	50	Bowler
R Ashwin	9.5	24.22	Bowler
S Badrinath	30.65	50	Batsman
L Balaji	3.66	26.09	Bowler
DE Bollinger	21	18.72	Batsman
DJ Bravo	25.1	22.47	Batsman
DL Chahar	10.83	28.2	Bowler
SM Curran	22	30.54	Batsman
MS Dhoni	40.53	50	Batsman
F du Plessis	35.33	50	Batsman
RD Gaikwad	39.36	50	Batsman
MS Gony	12.5	30.73	Bowler
Harbhajan Singh	7.5	25.3	Bowler

Research inquiries:

(i) **Classification of players:**
What is the use of FCM-σ clustering for categorizing players into three distinct categories based on their batting and bowling averages: Our system identifies players into three separate groups, which are batsmen, bowling all-rounders, and bowlers.

(ii) **Determination of optimal bowling all-rounders:**
What are the profiles of players in our sample who perform well both as bowlers and batsmen, and how do they stack up versus other player types?

(iii) **Cluster analysis:**
How does each performance segment appear different through its bowling and batting statistics?
The discovered clusters show how much they match the player categories from step one.

(iv) **Performance analysis:**
By using FCM-σ clustering, we can find player performance patterns that we cannot see through simple data review.

Solution:

Step 1: Consider batting average = X1 and bowling average = X2

X1	24.96	19	9.5	30.65	3.66	21	25.1	10.83	22	40.53	35.33	39.36	12.5	7.5
X2	22.78	50	24.22	50	26.09	18.72	22.47	28.2	30.54	50	50	50	30.73	25.3

Step 2: Starting partition of the membership function into three clustering centres

Mu1	0.009	0.233	0.954	0.019	0.914	0.113	0.012	0.972	0.213	0.019	0.000	0.012	0.851	0.976
Mu2	0.003	0.566	0.006	0.957	0.017	0.020	0.004	0.004	0.070	0.953	0.996	0.969	0.025	0.004
Mu3	0.988	0.200	0.040	0.024	0.068	0.867	0.984	0.024	0.717	0.029	0.000	0.019	0.125	0.020

Step 3: Take $m = 2$

Mu1^2	Mu2^2	Mu3^2	X1*Mu1^2	X2*Mu1^2	X1*Mu2^2	X2*Mu2^2	X1*Mu3^2	X2*Mu3^2
8.91E-05	9.32E-06	0.975172	0.002224	0.002029637	0.000233	0.000212	24.34029	22.21441982
0.05444	0.320807	0.040111	1.034363	2.722008372	6.095334	16.04035	0.762111	2.005554055
0.909592	3.58E-05	0.001623	8.641121	22.03031173	0.00034	0.000868	0.01542	0.039312236
0.00035	0.91563	0.000596	0.010713	0.01747694	28.06405	45.78148	0.018276	0.029813494
0.83604	0.000306	0.004645	3.059906	21.8122833	0.00112	0.007985	0.017	0.121184326
0.012696	0.000417	0.75152	0.266622	0.237674597	0.008756	0.007806	15.78191	14.06844884
0.000152	1.62E-05	0.96755	0.003817	0.003417342	0.000407	0.000364	24.28549	21.74083743
0.944111	1.74E-05	0.000585	10.22472	26.62392766	0.000188	0.00049	0.006331	0.01648642
0.045331	0.004912	0.514096	0.997284	1.384412102	0.108059	0.150005	11.31012	15.70050067
0.000353	0.907604	0.000814	0.014316	0.017660613	36.78521	45.38022	0.032974	0.040678988
0	0.991394	0	0	0	35.02595	49.56971	0	0
0.000154	0.938844	0.000348	0.006059	0.00769727	36.95291	46.94221	0.013694	0.017396339
0.723552	0.000611	0.01554	9.044404	22.23476191	0.007638	0.018778	0.194256	0.47755799
0.951905	1.5E-05	0.000419	7.139289	24.08320183	0.000112	0.000379	0.003144	0.010607031
4.478766	**4.080619**	**3.273019**	**40.44484**	**121.1768633**	**143.0503**	**203.9009**	**76.78103**	**76.48279764**

Step 4: Cluster centers

v1	08.9010	26.8672
v2	34.5691	49.9402
v3	23.4399	23.2381

Step 5: Membership values

NMu1	0.0080296	0.1699402	0.9513938	0.0099981	0.90353	0.0962779	0.0103518	0.9599718	0.1890907	0.0149379	0.0003174	0.0102651	0.8146891	0.9765935
NMu2	0.0039483	0.6638905	0.0080747	0.9751823	0.02473415	0.0262732	0.0051344	0.0075698	0.0973416	0.9593681	0.9991667	0.9722437	0.0394268	0.0047306
NMu3	0.9880219	0.1661692	0.0405314	0.0148195	0.071730287	0.8774487	0.9845136	0.0324582	0.7135676	0.0256938	0.0005158	0.0174911	0.1458840	0.0186758

Step 6: $e = 0.0001$

e1	1.2345e-05	1.2802e-05	1.0908e-05	2.4357e-06	1.5862e-05	1.2342e-05	1.1061e-05	1.4344e-05	1.3685e-05	1.5236e-06	2.0403e-09	1.1061e-05	2.4724e-05	1.5305e-05
e2	1.2825e-05	1.2845e-05	1.2797e-05	2.3711e-05	1.3567e-05	1.2406e-05	1.3758e-05	1.6759e-05	1.2974e-05	1.3806e-05	1.5774e-05	1.5746e-05	1.2935e-05	1.2759e-05
e3	3.7487e-06	1.3903e-05	1.2697e-05	1.4705e-05	1.2668e-05	2.7152e-06	3.8749e-06	1.4706e-05	1.3326e–05	1.6706e-05	3.0403e-09	1.0961e-05	1.2945e-05	1.2759e-05

Example 2

An organization has warehouses for life-saving drugs at six different locations, whose coordinates are given in the following table. The organization is in the process of locating one more critical central warehouse, which will distribute drugs to all the existing warehouses on emergency request. Find the number of optimum locations of the new facility (warehouse) based on the FCM-σ clustering concept.

Table 4.2: Thirty warehouses whose coordinates are summarized.

Existing warehouse number	Coordinates of centroids
1	(2.5,3.5)
2	(2.7,3.6)
3	(2.2,3.2)
4	(3.0,3.0)
5	(2.7,3.1)
6	(2.3,3.5)
7	(3.1,3.0)
8	(3.3,2.9)
9	(3.0,3.5)
10	(2.9, 2.9)
11	(6.0,7.2)
12	(6.5,7.0)
13	(5.9,6.9)
14	(6.3,7.3)
15	(5.9,7.0)
16	(6.8,7.0)
17	(6.5,6.9)
18	(6.2,7.1)
19	(7.2,7.2)
20	(6.6,7.1)
21	(11.1,13.5)
22	(12.2,12.5)
23	(11.5,11.9)

Solution:

The data are shown in Table 4.2. Assume that we want to determine a fuzzy pseudo-partition with three clusters (i.e. $c = 3$). Assume further that we choose $m = 2$. $||\,.\,||$ is the Euclidean distance, and the initial fuzzy pseudo-partition is $U^{(0)} = [U_1, U_2, U_3]$ with results provided in Tables 4.3–4.7.

Table 4.3: Cluster centres and corresponding membership function values of matrix *U*.

Input data		Membership function for fuzzy clustering		
X_1	X_2	U_1	U_2	U_3
2.5	3.5	.9	.05	.05
2.7	3.6	.89	.055	.055
2.2	3.2	.88	.06	.06
3.0	3.0	.87	.065	.065
2.7	3.1	.86	.07	.07
2.3	3.5	.84	.09	.07
3.1	3.0	.81	.11	.08
3.3	2.9	.8	.1	.1
3.0	3.5	.79	.11	.1
2.9	2.9	.75	.15	.1
6.0	7.2	.05	.9	.05
6.5	7.0	.055	.89	.055
5.9	6.9	.06	.88	.06
6.3	7.3	.065	.87	.065
5.9	7.0	.07	.86	.07
6.8	7.0	.09	.84	.07
6.5	6.9	.11	.81	.08
6.2	7.1	.1	.8	.1
7.2	7.2	.11	.79	.1
6.6	7.1	.15	.75	.1
11.1	13.5	.05	.05	.9
12.2	12.5	.055	.055	.89
11.5	11.9	.06	.06	.88
13.2	13.3	.065	.065	.87
11.5	12.5	.07	.07	.86
12.0	15.0	.07	.09	.84
13.2	13.2	.08	.11	.81
11.2	11.9	.1	.1	.8
12.9	13.7	.1	.11	.79
13.9	13.0	.1	.15	.75

Iteration 1:

Table 4.4: Cluster centres V and the corresponding membership function values of matrix U (Iteration 1).

Cluster centres		V_1	V_2	V_3
		(2.7717, 3.2218)	**(6.3887, 7.0712)**	**(12.2703, 13.0430)**
Average values		**(2.77, 3.22)**	**(6.38, 7.07)**	**(12.27, 13.04)**
Data points		U_1	U_2	U_3
2.5	3.5	0.9924	0.0053	0.0023
2.7	3.6	0.9921	0.0056	0.0023
2.2	3.2	0.9857	0.0097	0.0046
3.0	3.0	0.9949	0.0035	0.0015
2.7	3.1	0.9990	0.0007	0.0003
2.3	3.5	0.9858	0.0098	0.0044
3.1	3.0	0.9920	0.0056	0.0024
3.3	2.9	0.9805	0.0136	0.0058
3.0	3.5	0.9927	0.0052	0.0021
2.9	2.9	0.9943	0.0039	0.0018
6.0	7.2	0.0064	0.9870	0.0066
6.5	7.0	0.0006	0.9986	0.0007
5.9	6.9	0.0115	0.9787	0.0098
6.3	7.3	0.0021	0.9953	0.0025
5.9	7.0	0.0102	0.9808	0.0091
6.8	7.0	0.0058	0.9867	0.0075
6.5	6.9	0.0015	0.9967	0.0017
6.2	7.1	0.0014	0.9971	0.0015
7.2	7.2	0.0185	0.9503	0.0313
6.6	7.1	0.0016	0.9965	0.0020
11.1	13.5	0.0031	0.0084	0.9885
12.2	12.5	0.0006	0.0016	0.9978
11.5	11.9	0.0043	0.0129	0.9827
13.2	13.3	0.0015	0.0037	0.9947
11.5	12.5	0.0019	0.0054	0.9927
12.0	15.0	0.0060	0.0139	0.9801
13.2	13.2	0.0015	0.0036	0.9949
11.2	11.9	0.0057	0.0177	0.9766
12.9	13.7	0.0014	0.0033	0.9954
13.9	13.0	0.0042	0.0098	0.9860

Iteration 2:

Table 4.5: Cluster centres *V* and the corresponding membership function values of matrix *U* (Iteration 2).

Cluster centres		V_1	V_2	V_3
		(2.7706, 3.2204)	(6.3861, 7.0698)	(12.2748,13.0499)
Average values		(2.77, 3.22)	(6.39, 7.07)	(12.27,13.05)
Data points		U_1	U_2	U_3
2.5	3.5	0.9924	0.0052	0.0024
2.7	3.6	0.9920	0.0056	0.0024
2.2	3.2	0.9856	0.0096	0.0048
3.0	3.0	0.9949	0.0035	0.0016
2.7	3.1	0.9991	0.0006	0.0003
2.3	3.5	0.9857	0.0097	0.0046
3.1	3.0	0.9920	0.0055	0.0025
3.3	2.9	0.9805	0.0135	0.0060
3.0	3.5	0.9926	0.0052	0.0022
2.9	2.9	0.9943	0.0039	0.0018
6.0	7.2	0.0064	0.9868	0.0068
6.5	7.0	0.0007	0.9986	0.0008
5.9	6.9	0.0115	0.9784	0.0101
6.3	7.3	0.0021	0.9952	0.0027
5.9	7.0	0.0101	0.9805	0.0093
6.8	7.0	0.0059	0.9862	0.0080
6.5	6.9	0.0016	0.9966	0.0018
6.2	7.1	0.0014	0.9971	0.0015
7.2	7.2	0.0188	0.9485	0.0328
6.6	7.1	0.0016	0.9963	0.0021
11.1	13.5	0.0030	0.0081	0.9889
12.2	12.5	0.0006	0.0016	0.9978
11.5	11.9	0.0042	0.0125	0.9832
13.2	13.3	0.0015	0.0035	0.9950
11.5	12.5	0.0019	0.0053	0.9928
12.0	15.0	0.0058	0.0132	0.9810
13.2	13.2	0.0014	0.0034	0.9952
11.2	11.9	0.0056	0.0171	0.9773
12.9	13.7	0.0013	0.0031	0.9956
13.9	13.0	0.0040	0.0093	0.9866

Due to space and readability considerations, it is impractical to present the complete set of iterative results. Therefore, the following tables provide a summary of the membership values and cluster centres obtained during the penultimate (i.e. 11th) and final (i.e. 12th) iterations, which effectively illustrate the convergence behaviour of the algorithm.

Iteration 11:

Table 4.6: Cluster centres *V* and the corresponding membership function values of matrix *U* (Iteration 11).

Cluster centres		V_1	V_2	V_3
		(2.7706, 3.2204)	**(6.3858, 7.0697)**	**(12.2746, 13.0499)**
Average values		**(2.77, 3.22)**	**(6.39, 7.07)**	**(12.27, 13.05)**
Data points		U_1	U_2	U_3
2.5	3.5	0.9924	0.0052	0.0024
2.7	3.6	0.9920	0.0056	0.0024
2.2	3.2	0.9857	0.0096	0.0048
3.0	3.0	0.9949	0.0035	0.0016
2.7	3.1	0.9991	0.0006	0.0003
2.3	3.5	0.9857	0.0097	0.0046
3.1	3.0	0.9920	0.0055	0.0025
3.3	2.9	0.9805	0.0135	0.0060
3.0	3.5	0.9927	0.0052	0.0022
2.9	2.9	0.9943	0.0039	0.0018
6.0	7.2	0.0064	0.9868	0.0068
6.5	7.0	0.0007	0.9986	0.0008
5.9	6.9	0.0115	0.9784	0.0101
6.3	7.3	0.0021	0.9952	0.0027
5.9	7.0	0.0101	0.9805	0.0093
6.8	7.0	0.0059	0.9861	0.0080
6.5	6.9	0.0016	0.9966	0.0018
6.2	7.1	0.0014	0.9971	0.0015
7.2	7.2	0.0188	0.9483	0.0329
6.6	7.1	0.0016	0.9963	0.0021
11.1	13.5	0.0030	0.0081	0.9889
12.2	12.5	0.0006	0.0016	0.9978
11.5	11.9	0.0042	0.0125	0.9833
13.2	13.3	0.0015	0.0035	0.9950
11.5	12.5	0.0019	0.0053	0.9929
12.0	15.0	0.0058	0.0132	0.9811
13.2	13.2	0.0014	0.0034	0.9952
11.2	11.9	0.0056	0.0170	0.9774
12.9	13.7	0.0013	0.0031	0.9956
13.9	13.0	0.0040	0.0093	0.9867

Iteration 12:

Table 4.7: Cluster centres *V* and the corresponding membership function values of matrix *U* (Iteration 12).

Cluster centres		V_1	V_2	V_3
		(2.27706, 3.2204)	**(6.3858, 7.0697)**	**(12.2746, 13.0499)**
Average values		**(2.28, 3.22)**	**(6.39, 7.07)**	**(12.27,13.05)**
Data points		U_1	U_2	U_3
2.5	3.5	0.9924	0.0052	0.0024
2.7	3.6	0.9920	0.0056	0.0024
2.2	3.2	0.9856	0.0096	0.0048
3.0	3.0	0.9949	0.0035	0.0016
2.7	3.1	0.9990	0.0006	0.0003
2.3	3.5	0.9857	0.0097	0.0046
3.1	3.0	0.9920	0.0055	0.0025
3.3	2.9	0.9804	0.0135	0.0060
3.0	3.5	0.9926	0.0052	0.0022
2.9	2.9	0.9943	0.0039	0.0018
6.0	7.2	0.0064	0.9868	0.0068
6.5	7.0	0.0007	0.9986	0.0008
5.9	6.9	0.0115	0.9784	0.0101
6.3	7.3	0.0021	0.9952	0.0027
5.9	7.0	0.0101	0.9805	0.0093
6.8	7.0	0.0059	0.9861	0.0080
6.5	6.9	0.0016	0.9966	0.0018
6.2	7.1	0.0014	0.9971	0.0015
7.2	7.2	0.0188	0.9483	0.0329
6.6	7.1	0.0016	0.9963	0.0021
11.1	13.5	0.0030	0.0081	0.9889
12.2	12.5	0.0006	0.0016	0.9978
11.5	11.9	0.0042	0.0125	0.9833
13.2	13.3	0.0015	0.0035	0.9950
11.5	12.5	0.0019	0.0053	0.9929
12.0	15.0	0.0058	0.0132	0.9811
13.2	13.2	0.0014	0.0034	0.9952
11.2	11.9	0.0056	0.0170	0.9774
12.9	13.7	0.0013	0.0031	0.9956
13.9	13.0	0.0040	0.0093	0.9867

Then, the algorithm stops after 12 iterations, and we obtain the pseudo-partition defined in Table 4.7. The three cluster centres are $V_1 = (2.27706, 3.2204)$, $V_2 = (6.3858, 7.0697)$, and $V_3 = (12.2746, 13.0499)$, and $J_{FCM-\sigma} = 20.8168$.

4.6 Problem set

Problem 1

This study records the daily **sleep hours** and **study hours** of five students. However, one student reports unusually high sleep and low study hours, possibly due to illness or error. The **FCM-σ** technique is applied to cluster students into:
- **Focused learners** (less sleep, more study)
- **Balanced learners**
- **Less studious** (more sleep, less study)

FCM-σ improves results by reducing the impact of noisy or outlier entries.

Table 1: Students' daily habits.

Student	Sleep hours	Study hours
A	6	5
B	5.5	6
C	8	2
D	6.5	5
E	12	0 (noisy)

Research inquiries:
(i) **Handling outliers in FCM-σ:** How does FCM-σ reduce the influence of Student E on the cluster centres compared to standard FCM?
(ii) **Cluster profiles:** What do the three cluster centres represent in terms of sleep and study balance?
(iii) **Membership comparison:** How does the membership of Student E differ between FCM and FCM-σ?
(iv) **Improved real-life clustering:** How can FCM-σ be more reliable than standard FCM for behaviour-based datasets?

Problem 2

We analyse the **fitness profiles** of five individuals using three parameters: **Steps per Day**, **Calories Burned**, and **Water Intake (litres)**. One person shows an irregular pattern, likely due to faulty recording. FCM-σ is used to cluster into:
- **Active and healthy**
- **Low activity**
- **Balanced** lifestyle groups

Table 2: Fitness tracker data.

Person	Steps	Calories burned	Water intake
P1	8,000	280	2.0
P2	8,200	270	2.2
P3	1,500	90	1.5
P4	7,800	300	2.1
P5	900	800	0.5 (noisy)

Research inquiries:

(i) **Cluster robustness with FCM-σ:** How is the influence of noisy data from P5 reduced in cluster formation?

(ii) **Interpretation of cluster** centres**:** What kinds of health profiles emerge from the clustering?

(iii) **Membership stabilization:** How does FCM-σ help in preventing drastic shifts in cluster centres caused by anomalies?

(iv) **Wellness applications:** Where can FCM-σ be used in real-world fitness or wellness analytics?

Problem 3

This study evaluates employees based on **Task Completion Rate**, **Teamwork Score**, **Error Rate**, and **Customer Feedback Score**. The dataset includes one inconsistent performer. Using FCM-σ, employees are clustered into:

- **Reliable performers**
- **Inconsistent workers**
- **Underperformers**

Table 3: Employee performance metrics.

Employee	Tasks done (%)	Teamwork	Errors	Customer feedback
Emp_1	95	80	2	4.5
Emp_2	92	85	1	4.7
Emp_3	60	50	10	3.0
Emp_4	90	78	2	4.3
Emp_5	15	20	1	4.9 (noisy)

Research inquiries:

(i) **Effect of robustness on membership weights:** Does FCM-σ correctly prevent Emp_5's anomalous feedback score from shifting the cluster?

(ii) **Cluster** centre **meaning:** What traits do the resulting clusters represent (consistent vs. inconsistent performers)?

(iii) **Noise resistance comparison:** Compare how FCM and FCM-σ handle the noise introduced by Emp_5.

(iv) **HR use case:** How can organizations use FCM-σ for fair employee evaluation?

Chapter 5
PCM-σ (robust possibilistic *C*-means)

5.1 Introduction

PCB-σ (robust possibilistic *C*-means) can be considered as a modification of the simple possibilistic *C*-means (PCM) where its purpose is to enhance the robustness of the standard algorithm especially in cases of noisy data or the presence of outliers. Typicality values of traditional PCM convey how well each data point reflects a cluster, in order to avoid sensitivity to noise. Nevertheless, PCM remains prone to coincident clusters (two or more clusters are collapsed on to one centre), as well as instability when the distribution of data is intricate.

These issues become alleviated in PCM-σ since it adds an adaptive scale parameter, σ, to the clustering procedure. This value of regularized covariance is representative of the dispersion or variance of any cluster and it changes dynamically upon iterations. In this way, PCM-σ has a better grip on the effect of the far-flung (perhaps noisy) data points on the cluster centres that results in more resistant and significant clustering outcomes. Due to this, PCM-σ is particularly useful with real-world data whose data can be characterized by much noise data, irregular cluster shapes, or outliers.

The PCM distance calculation method is based entirely on the standard geometric distances between data points and their respective centroids. It doesn't care how close together the data points are as long as they're in the same cluster. Consequently, PCM's efficacy might diminish when confronted with unevenly dispersed data points or clusters that deviate from perfect sphericity [33, 47, 48]. The enhanced distance measure differs from the Mahalanobis distance in that it does not scale distances according to inter-feature correlation but instead standardizes them based on the dispersion of data points around a cluster's prototype. In order to find C clusters, PCM-σ tries to minimize this objective:

$$J_{\text{PCM}-\sigma} = \sum_{j=1}^{N} \sum_{I=1}^{C} t_{ij}^{m} \frac{\|x_j - v_i\|^2}{\sigma_i^2} + \sum_{i=1}^{C} Y_i \sum_{j=1}^{N} (1 - t_{ij})^m$$

where

$$\sigma_i^2 = \frac{\sum_{j=1}^{N} t_{ij}^{m} \|x_j - v_i\|^2}{\sum_{j=1}^{N} t_{ij}^{m}}$$

And by partially differentiating the above objective function with respect to t_{ij} and v_i, and subsequently equating them to zero, we obtain the values of typicality and cluster prototype, respectively:

https://doi.org/10.1515/9783111494272-005

$$t_{ij} = \frac{1}{1 + \left(\frac{\frac{\|x_j - v_i\|^2}{\sigma_i^2}}{Y_i} \right)^{\frac{1}{m-1}}}$$

$$1 \le i \le C; 1 \le j \le N$$

$$\mathrm{v_i} = \frac{\sum_{j=1}^{N} t_{ij}^{m} x_j}{\sum_{j=1}^{N} t_{ij}^{m}}, \quad 1 \le i \le C$$

Similarly, we can find the value of Y_i, such that

$$Y_i = K \frac{\sum_{j=1}^{N} t_{ij}^{m} \frac{\|x_j - v_i\|^2}{\sigma_i^2}}{\sum_{j=1}^{N} t_{ij}^{m}}, \quad 1 \le i \le C$$

5.2 Mathematical derivation for robust PCM

Derivation for possibilistic membership t_{ij}

For finding possibilistic membership, compute the partial derivative of $J_{\text{PCM}-\sigma}$ with respect to t_{ij}:

$$\frac{\partial J_{\text{PCM}-\sigma}}{\partial t_{ij}} = m t_{ij}^{m-1} \frac{\|x_j - v_i\|^2}{\sigma_i^2} - m Y_i (1 - t_{ij})^{m-1}$$

Set it to zero:

$$\Rightarrow m t_{ij}^{m-1} \frac{\|x_j - v_i\|^2}{\sigma_i^2} - m Y_i (1 - t_{ij})^{m-1} = 0$$

$$\Rightarrow m t_{ij}^{m-1} \frac{\|x_j - v_i\|^2}{\sigma_i^2} = m Y_i (1 - t_{ij})^{m-1}$$

$$\Rightarrow \frac{t_{ij}^{m-1}}{(1 - t_{ij})^{m-1}} = \frac{Y_i}{\|x_j - v_i\|^2}$$

Now solve it for t_{ij}:

$$=> \left(\frac{t_{ij}}{1-t_{ij}}\right)^{m-1} = \frac{Y_i\sigma_i^2}{||x_j - v_i||^2}$$

$$=> \frac{t_{ij}}{1-t_{ij}} = \left(\frac{Y_i\sigma_i^2}{||x_j - v_i||^2}\right)^{\frac{1}{m-1}}$$

$$=> t_{ij} = \frac{1}{1+\left(\frac{||x_j - v_i||^2}{Y_i\sigma_i^2}\right)^{\frac{1}{m-1}}}$$

Derivation for cluster centre v_i

Now, for finding cluster centre, we take the derivative of $J_{\text{PCM}-\sigma}$ with respect to v_i and then set it to zero:

$$\frac{\partial J_{\text{PCM}-\sigma}}{\partial v_i} = \sum_{j=1}^{N} -2(t_{ij})^m \frac{(x_j - v_i)}{\sigma_i^2} = 0$$

$$=> \frac{-2}{\sigma_i^2}\sum_{i=1}^{C}(t_{ij})^m(x_j - v_i) = 0$$

$$=> \sum_{i=1}^{C}(t_{ij})^m(x_j - v_i) = 0$$

Solving for v_i we get

$$v_i = \frac{\sum_{j=1}^{N} t_{ij}^m x_j}{\sum_{j=1}^{N} t_{ij}^m}$$

Derivation for scale parameter Y_i

The scale parameter comes from empirical optimization, not purely calculus. It can't be calculated as v_i and t_{ij} as t_{ij} depends on Y_i. It is usually estimated using weighted mean square error. So,

$$Y_i = K\frac{\sum_{j=1}^{N} t_{ij}^m \frac{||x_j - v_i||^2}{\sigma_i^2}}{\sum_{j=1}^{N} t_{ij}^m}$$

Derivation of σ_i^2

For finding σ_i^2 partial derivative would not help. So, instead of relying on setting the derivative to zero we can recognize that the objective function can be seen as a weighted sum of squared distances. So, to minimize $J_{PCM-\sigma}$, the best of value of σ_i^2 will be the one that balances this weighted sum.

The optimal value of σ_i^2 can be obtained by recognizing that the scale σ_i^2 corresponding to a weighted average of square distance. Thus, we can directly write

$$\sigma_i^2 = \frac{\sum_{j=1}^{N} t_{ij}^m \left\| x_j - v_i \right\|^2}{\sum_{j=1}^{N} t_{ij}^m}$$

5.3 Python code

```
import numpy as np
import matplotlib.pyplot as plt

def pcm_sigma(X, C=3, m=2, max_iter=100, tol=1e-5, seed=42):
  np.random.seed(seed)
  N, d = X.shape

  # Step 1: Initialize random membership matrix (C x N)
  mu = np.random.rand(C, N)
  mu /= np.sum(mu, axis=0, keepdims=True)

  obj_values = []

  for iteration in range(max_iter):
    # Step 2: Compute cluster centers
    mu_m = mu ** m
    v = (mu_m @ X) / np.sum(mu_m, axis=1, keepdims=True)

    # Step 3: Compute distances (C x N)
    D = np.linalg.norm(X[None, :, :] - v[:, None, :], axis=2) +
    1e-8

    # Step 4: Compute σ_i
    sigma = np.sqrt(np.sum(mu_m * (D ** 2), axis=1) / np.sum(mu_m,
    axis=1))
```

```
# Step 5: Update membership values
mu_new = np.zeros_like(mu)
for j in range(N):
  for i in range(C):
    denom = np.sum(
      ((D[i, j]**2 / sigma[i]) / (D[:, j]**2 / sigma)) ** (1/
      (m-1))
    )
    mu_new[i, j] = 1 / (denom + 1e-8)

# Step 6: Compute objective function
J = np.sum(mu_new**m * (D**2) / sigma[:, None])
obj_values.append(J)

# Step 7: Check convergence
if np.max(np.abs(mu_new - mu)) < tol:
   print(f"Converged at iteration {iteration+1}")
   break

mu = mu_new
# Plot objective function
plt.figure(figsize=(6,4))
plt.plot(obj_values, marker='o')
plt.title("PCM-σ Objective Function over Iterations")
plt.xlabel("Iteration")
plt.ylabel("Objective Function")
plt.show()

# Plot clustering result
labels = np.argmax(mu, axis=0)
plt.figure(figsize=(6,6))
plt.scatter(X[:,0], X[:,1], c=labels, cmap='viridis',
alpha=0.6)
plt.scatter(v[:,0], v[:,1], color='red', marker='x', s=100,
label='Centers')
plt.title("PCM-σ Clustering Result")
plt.legend()
plt.show()

return v, mu, sigma, obj_values
```

```
        # Example usage
        np.random.seed(0)
        X = np.vstack([
          np.random.randn(50,2) + np.array([2,2]),
          np.random.randn(50,2) + np.array([-2,-2]),
          np.random.randn(50,2) + np.array([5,-3])
        ])
        centers, membership, sigma, obj = pcm_sigma(X, C=3)
        print("Cluster centers:\n", centers)
print("Sigma values:\n", sigma)
```

5.4 R code

```
pcm_sigma <- function(X, C=3, m=2, max_iter=100, tol=1e-5, seed=42) {
  set.seed(seed)
  N <- nrow(X)
  d <- ncol(X)

  # Initialize random membership matrix
  mu <- matrix(runif(C*N), nrow=C)
  mu <- mu / matrix(colSums(mu), nrow=C, ncol=N, byrow=TRUE)

  obj_values <- c()

  for (iter in 1:max_iter) {
   # Compute cluster centers
   mu_m <- mu^m
   v <- (mu_m %*% as.matrix(X)) / rowSums(mu_m)

  # Compute distances
  D <- matrix(0, nrow=C, ncol=N)
  for (i in 1:C) {
   diff <- sweep(X, 2, v[i,])
   D[i, ] <- sqrt(rowSums(diff^2)) + 1e-8
  }

  # Compute sigma
  sigma <- sqrt(rowSums(mu_m * D^2) / rowSums(mu_m))

  # Update membership values
  mu_new <- matrix(0, nrow=C, ncol=N)
```

```
for (j in 1:N) {
  for (i in 1:C) {
    ratios <- (D[i,j]^2 / sigma[i]) / (D[,j]^2 / sigma)
    denom <- sum(ratios^(1/(m-1)))
    mu_new[i,j] <- 1 / (denom + 1e-8)
  }
}

# Compute objective function
J <- sum(mu_new^m * (D^2) / sigma)
obj_values <- c(obj_values, J)

# Check convergence
if (max(abs(mu_new - mu)) < tol) {
  cat(sprintf("Converged at iteration %d\n", iter))
  break
}

  mu <- mu_new
}

# Plot objective
plot(obj_values, type="o", col="blue", xlab="Iteration",
ylab="Objective Function",
  main="PCM-σ Objective Function over Iterations")

# Plot clustering result
cluster_labels <- apply(mu, 2, which.max)
df <- data.frame(X1=X[,1], X2=X[,2], cluster=as.factor
(cluster_labels))
centers_df <- data.frame(v)
colnames(centers_df) <- c("X1", "X2")

library(ggplot2)
ggplot(df, aes(x=X1, y=X2, color=cluster)) +
 geom_point(size=2, alpha=0.7) +
 geom_point(data=centers_df, aes(x=X1, y=X2), color='red', size=4,
shape=4, stroke=1.5) +
 ggtitle("PCM-σ Clustering Result") +
 theme_minimal()
```

```
return(list(centers=v, membership=mu, sigma=sigma,
objective=obj_values))
}

# Example usage
set.seed(0)
X <- rbind(
 matrix(rnorm(100, mean=2), ncol=2),
 matrix(rnorm(100, mean=-2), ncol=2),
 matrix(rnorm(100, mean=4), ncol=2)
)
result <- pcm_sigma(X, C=3)
print("Cluster centers:")
print(result$centers)
print("Sigma values:")
print(result$sigma)
```

5.5 Solved examples

Example 1

The study investigates 15 CSK IPL players' stats including batting averages, over rates, and selection positions shown in Table 5.1. Employing the robust PCM clustering technique, we want to categorize the players into three unique clusters. Our analysis groups players into batters, all-rounders, and fast bowlers. Through this technique, we discover the strongest bowling all-rounder performers within our dataset.

Table 5.1: CSK team members' batting and bowling statistics along with their early designations.

Player	Batting average	Bowl average	Initial class
MM Ali	24.96	22.78	Batsman
S Anirudha	19	50	Bowler
R Ashwin	9.5	24.22	Bowler
S Badrinath	30.65	50	Batsman
L Balaji	3.66	26.09	Bowler
DE Bollinger	21	18.72	Batsman
DJ Bravo	25.1	22.47	Batsman
DL Chahar	10.83	28.2	Bowler
SM Curran	22	30.54	Batsman
MS Dhoni	40.53	50	Batsman

Table 5.1 (continued)

Player	Batting average	Bowl average	Initial class
F du Plessis	35.33	50	Batsman
RD Gaikwad	39.36	50	Batsman
MS Gony	12.5	30.73	Bowler
Harbhajan Singh	7.5	25.3	Bowler

Research inquiries:

(i) **Classification of players:**
What is the use of robust PCM clustering for categorizing players into three distinct categories based on their batting and bowling averages: Our system identifies players into three separate groups, which are batsmen, bowling all-rounders, and bowlers.

(ii) **Determination of optimal bowling all-rounders:**
What are the profiles of players in our sample who perform well both as bowlers and batsmen, and how do they stack up versus other player types?

(iii) **Cluster analysis:**
How does each performance segment appear different through its bowling and batting statistics?
The discovered clusters show how much they match the player categories from step one.

(iv) **Performance analysis:**
By using robust PCM clustering, we can find player performance patterns that we cannot see through simple data review.

Solution:

Step 1: Consider batting average = X1 and bowling average = X2.

X1	24.96	19	9.5	30.65	3.66	21	25.1	10.83	22	40.53	35.33	39.36	12.5	7.5
X2	22.78	50	24.22	50	26.09	18.72	22.47	28.2	30.54	50	50	50	30.73	25.3

Step 2: Starting partition of the membership function into three clustering centres.

Mu1	0.009	0.233	0.954	0.019	0.914	0.113	0.012	0.972	0.213	0.019	0.000	0.012	0.851	0.976
Mu2	0.003	0.566	0.006	0.957	0.017	0.020	0.004	0.004	0.070	0.953	0.996	0.969	0.025	0.004
Mu3	0.988	0.200	0.040	0.024	0.068	0.867	0.984	0.024	0.717	0.029	0.000	0.019	0.125	0.020

Step 3: Take $m = 2$.

Mu1^2	Mu2^2	Mu3^2	X1*Mu1^2	X2*Mu1^2	X1*Mu2^2	X2*Mu2^2	X1*Mu3^2	X2*Mu3^2
8.91E-05	9.32E-06	0.975172	0.002224	0.002029637	0.000233	0.000212	24.34029	22.21441982
0.05444	0.320807	0.040111	1.034363	2.722008372	6.095334	16.04035	0.762111	2.005554055
0.909592	3.58E-05	0.001623	8.641121	22.03031173	0.00034	0.000868	0.01542	0.039312236
0.00035	0.91563	0.000596	0.010713	0.01747694	28.06405	45.78148	0.018276	0.029813494
0.83604	0.000306	0.004645	3.059906	21.8122833	0.00112	0.007985	0.017	0.121184326
0.012696	0.000417	0.75152	0.266622	0.237674597	0.008756	0.007806	15.78191	14.06844884
0.000152	1.62E-05	0.96755	0.003817	0.003417342	0.000407	0.000364	24.28549	21.74083743
0.944111	1.74E-05	0.000585	10.22472	26.62392766	0.000188	0.00049	0.006331	0.01648642
0.045331	0.004912	0.514096	0.997284	1.384412102	0.108059	0.150005	11.31012	15.70050067
0.000353	0.907604	0.000814	0.014316	0.017660613	36.78521	45.38022	0.032974	0.040678988
0	0.991394	0	0	0	35.02595	49.56971	0	0
0.000154	0.938844	0.000348	0.006059	0.00769727	36.95291	46.94221	0.013694	0.017396339
0.723552	0.000611	0.01554	9.044404	22.23476191	0.007638	0.018778	0.194256	0.47755799
0.951905	1.5E-05	0.000419	7.139289	24.08320183	0.000112	0.000379	0.003144	0.010607031
4.478766	**4.080619**	**3.273019**	**40.44484**	**121.1768633**	**143.0503**	**203.9009**	**76.78103**	**76.48279764**

Step 4: Cluster centers

v1	7.52	25.3
v2	35.33	50
v3	24.96	22.78

Step 5: Membership values

NMu1	0.00	0.00	0.00	0.00	0.00	0.00	0.00	0.00	0.00	0.00	0.00	0.00	0.00	0.5025
NMu2	0.00	0.00	0.00	0.00	0.00	0.00	0.00	0.00	0.00	0.00	0.5011	0.00	0.00	0.00
NMu3	0.5017	0.00	0.00	0.00	0.00	0.00	0.00	0.00	0.00	0.00	0.00	0.00	0.00	0.00

Step 6: $e = 0.0001$.

e1	1.2345e-05	1.2802e-05	1.0908e-05	2.4357e-06	1.5862e-05	1.2342e-05	1.1061e-05	1.4344e-05	1.3685e-05	1.5236e-06	2.0403e-09	1.1061e-05	2.4724e-05	1.5305e-05
e2	1.2825e-05	1.2845e-05	1.2797e-05	2.3711e-05	1.3567e-05	1.2406e-05	1.3758e-05	1.6759e-05	1.2974e-05	1.3806e-05	1.5774e-05	1.5746e-05	1.2935e-05	1.2759e-05
e3	3.7487e-06	1.3903e-05	1.2697e-05	1.4705e-05	1.2668e-05	2.7152e-06	3.8749e-06	1.4706e-05	1.3326e–05	1.6706e-05	3.0403e-09	1.0961e-05	1.2945e-05	1.2759e-05

Example 2

An organization has warehouses for life-saving drugs at six different locations, whose coordinates are given in the following table. The organization is in the process of locating one more critical central warehouse, which will distribute drugs to all the existing warehouses on emergency requests. Find the number of optimum locations of the new facility (warehouse) based on the robust PCM clustering concept.

Table 5.2: Thirty warehouses whose coordinates are summarized.

Existing warehouse number	Coordinates of centroids
1	(2.5,3.5)
2	(2.7,3.6)
3	(2.2,3.2)
4	(3.0,3.0)
5	(2.7,3.1)
6	(2.3,3.5)
7	(3.1,3.0)
8	(3.3,2.9)
9	(3.0,3.5)
10	(2.9, 2.9)
11	(6.0,7.2)
12	(6.5,7.0)
13	(5.9,6.9)
14	(6.3,7.3)
15	(5.9,7.0)
16	(6.8,7.0)
17	(6.5,6.9)
18	(6.2,7.1)
19	(7.2,7.2)
20	(6.6,7.1)
21	(11.1,13.5)
22	(12.2,12.5)
23	(11.5,11.9)

Solution:

The data are shown in Table 5.2. Assume that we want to determine a fuzzy pseudo-partition with three clusters (i.e. $c = 3$). Assume further that we choose $m = 2$. $||\ .\ ||$ is the Euclidean distance, and the initial fuzzy pseudo-partition is $U^{(0)} = [U_1, U_2, U_3]$. Table 5.3–5.6 shows the results obtained at different iterations.

Table 5.3: Cluster centres and corresponding membership function values of matrix U.

Input data		**Membership function for fuzzy clustering**		
X_1	X_2	U_1	U_2	U_3
2.5	3.5	.9	.05	.05
2.7	3.6	.89	.055	.055
2.2	3.2	.88	.06	.06
3.0	3.0	.87	.065	.065
2.7	3.1	.86	.07	.07
2.3	3.5	.84	.09	.07
3.1	3.0	.81	.11	.08
3.3	2.9	.8	.1	.1
3.0	3.5	.79	.11	.1
2.9	2.9	.75	.15	.1
6.0	7.2	.05	.9	.05
6.5	7.0	.055	.89	.055
5.9	6.9	.06	.88	.06
6.3	7.3	.065	.87	.065
5.9	7.0	.07	.86	.07
6.8	7.0	.09	.84	.07
6.5	6.9	.11	.81	.08
6.2	7.1	.1	.8	.1
7.2	7.2	.11	.79	.1
6.6	7.1	.15	.75	.1
11.1	13.5	.05	.05	.9
12.2	12.5	.055	.055	.89
11.5	11.9	.06	.06	.88
13.2	13.3	.065	.065	.87
11.5	12.5	.07	.07	.86
12.0	15.0	.07	.09	.84
13.2	13.2	.08	.11	.81
11.2	11.9	.1	.1	.8
12.9	13.7	.1	.11	.79
13.9	13.0	.1	.15	.75

Iteration 1:

Table 5.4: Cluster centres V and the corresponding membership function values of matrix U (Iteration 1).

Cluster centres		V_1	V_2	V_3
		(2.8723, 3.3594)	**(6.3981 7.0939)**	**(12.0856 12.9097)**
Average values		**(2.87, 3.36)**	**(6.39, 7.10)**	**(12.09, 12.91)**
Data points		U_1	U_2	U_3
2.5	3.5	0.9289	0.0471	0.0197
2.7	3.6	0.9594	0.0509	0.0204
2.2	3.2	0.8126	0.0406	0.0186
3.0	3.0	0.9343	0.0468	0.0197
2.7	3.1	0.9552	0.0448	0.0193
2.3	3.5	0.8563	0.0446	0.0193
3.1	3.0	0.9196	0.0478	0.0199
3.3	2.9	0.8401	0.0486	0.0201
3.0	3.5	0.9829	0.0537	0.0208
2.9	2.9	0.9072	0.0445	0.0193
6.0	7.2	0.0778	0.8911	0.0496
6.5	7.0	0.0727	0.8294	0.0521
5.9	6.9	0.0872	0.9638	0.0466
6.3	7.3	0.0705	0.8439	0.0530
5.9	7.0	0.0845	0.8907	0.0473
6.8	7.0	0.0673	0.9666	0.0546
6.5	6.9	0.0745	0.9725	0.0512
6.2	7.1	0.0763	0.6797	0.0504
7.2	7.2	0.0582	0.9714	0.0605
6.6	7.1	0.0691	0.0215	0.0538
11.1	13.5	0.0119	0.0216	0.7335
12.2	12.5	0.0119	0.9863	0.9526
11.5	11.9	0.0138	0.0275	0.7273
13.2	13.3	0.0099	0.0161	0.7227
11.5	12.5	0.0129	0.0245	0.8767
12.0	15.0	0.0094	0.0146	0.4536
13.2	13.2	0.0101	0.0163	0.7326
11.2	11.9	0.0143	0.0292	0.6683
12.9	13.7	0.0098	0.0159	0.7383
13.9	13.0	0.0095	0.0150	0.5241

Iteration 2:

Table 5.5: Cluster centres V and the corresponding membership function values of matrix U (Iteration 2).

Cluster centres		V_1	V_2	V_3
		(2.8069 3.2561)	**(6.3648 7.0607)**	**(12.1485 12.8523)**
Average values		**(2.81, 3.26)**	**(6.37, 7.06)**	**(12.15,12.85)**
Data points		U_1	U_2	U_3
2.5	3.5	0.7113	0.0088	0.0095
2.7	3.6	0.7448	0.0096	0.0098
2.2	3.2	0.5048	0.0076	0.0089
3.0	3.0	0.7863	0.0088	0.0095
2.7	3.1	0.9137	0.0084	0.0093
2.3	3.5	0.5447	0.0084	0.0093
3.1	3.0	0.7142	0.0090	0.0096
3.3	2.9	0.5057	0.0091	0.0097
3.0	3.5	0.7963	0.0101	0.0100
2.9	2.9	0.7365	0.0083	0.0093
6.0	7.2	0.0145	0.6173	0.0242
6.5	7.0	0.0135	0.9180	0.0255
5.9	6.9	0.0163	0.5042	0.0227
6.3	7.3	0.0131	0.8001	0.0260
5.9	7.0	0.0158	0.5282	0.0231
6.8	7.0	0.0125	0.5601	0.0268
6.5	6.9	0.0139	0.8479	0.0251
6.2	7.1	0.0142	0.8955	0.0247
7.2	7.2	0.0107	0.2554	0.0298
6.6	7.1	0.0128	0.8121	0.0264
11.1	13.5	0.0022	0.0038	0.5329
12.2	12.5	0.0022	0.0038	0.9318
11.5	11.9	0.0025	0.0049	0.5662
13.2	13.3	0.0018	0.0029	0.5703
11.5	12.5	0.0023	0.0044	0.7608
12.0	15.0	0.0017	0.0026	0.2722
13.2	13.2	0.0018	0.0029	0.5856
11.2	11.9	0.0026	0.0052	0.4896
12.9	13.7	0.0018	0.0028	0.5746
13.9	13.0	0.0017	0.0027	0.3594

Due to space and readability considerations, it is impractical to present the complete set of iterative results. Therefore, the following tables provide a summary of the membership values and cluster centres obtained during the penultimate (i.e. 35th) and final (i.e. 36th) iterations, which effectively illustrate the convergence behaviour of the algorithm.

Iteration 35:

Table 5.6: Cluster centres V and the corresponding membership function values of matrix U (Iteration 35).

Cluster centres		V_1	V_2	V_3
		(3, 3)	**(6.5, 7.00)**	**(11.5, 11.9)**
Average values		**(3, 3)**	**(6.5, 7.00)**	**(11.5, 11.9)**
Data points		U_1	U_2	U_3
2.5	3.5	0.00	0.00	0.00
2.7	3.6	0.00	0.00	0.00
2.2	3.2	0.00	0.00	0.00
3.0	3.0	0.5151	0.00	0.00
2.7	3.1	0.00	0.00	0.00
2.3	3.5	0.00	0.00	0.00
3.1	3.0	0.00	0.00	0.00
3.3	2.9	0.00	0.00	0.00
3.0	3.5	0.00	0.00	0.00
2.9	2.9	0.00	0.00	0.00
6.0	7.2	0.00	0.00	0.00
6.5	7.0	0.00	0.5069	0.00
5.9	6.9	0.00	0.00	0.00
6.3	7.3	0.00	0.00	0.00
5.9	7.0	0.00	0.00	0.00
6.8	7.0	0.00	0.00	0.00
6.5	6.9	0.00	0.00	0.00
6.2	7.1	0.00	0.00	0.00
7.2	7.2	0.00	0.00	0.00
6.6	7.1	0.00	0.00	0.00
11.1	13.5	0.00	0.00	0.00
12.2	12.5	0.00	0.00	0.00
11.5	11.9	0.00	0.00	0.5143
13.2	13.3	0.00	0.00	0.00
11.5	12.5	0.00	0.00	0.00
12.0	15.0	0.00	0.00	0.00
13.2	13.2	0.00	0.00	0.00
11.2	11.9	0.00	0.00	0.00
12.9	13.7	0.00	0.00	0.00
13.9	13.0	0.00	0.00	0.00

Iteration 36:

Table 5.7: Cluster centres V and the corresponding membership function values of matrix U (Iteration 36).

Cluster centres		V_1	V_2	V_3
		(3, 3)	(6.5, 7.00)	(11.5, 11.9)
Average values		(3, 3)	(6.5, 7.00)	(11.5, 11.9)
Data points		U_1	U_2	U_3
2.5	3.5	0.00	0.00	0.00
2.7	3.6	0.00	0.00	0.00
2.2	3.2	0.00	0.00	0.00
3.0	3.0	0.5151	0.00	0.00
2.7	3.1	0.00	0.00	0.00
2.3	3.5	0.00	0.00	0.00
3.1	3.0	0.00	0.00	0.00
3.3	2.9	0.00	0.00	0.00
3.0	3.5	0.00	0.00	0.00
2.9	2.9	0.00	0.00	0.00
6.0	7.2	0.00	0.00	0.00
6.5	7.0	0.00	0.5069	0.00
5.9	6.9	0.00	0.00	0.00
6.3	7.3	0.00	0.00	0.00
5.9	7.0	0.00	0.00	0.00
6.8	7.0	0.00	0.00	0.00
6.5	6.9	0.00	0.00	0.00
6.2	7.1	0.00	0.00	0.00
7.2	7.2	0.00	0.00	0.00
6.6	7.1	0.00	0.00	0.00
11.1	13.5	0.00	0.00	0.00
12.2	12.5	0.00	0.00	0.00
11.5	11.9	0.00	0.00	0.5037
13.2	13.3	0.00	0.00	0.00
11.5	12.5	0.00	0.00	0.00
12.0	15.0	0.00	0.00	0.00
13.2	13.2	0.00	0.00	0.00
11.2	11.9	0.00	0.00	0.00
12.9	13.7	0.00	0.00	0.00
13.9	13.0	0.00	0.00	0.00

Then, the algorithm stops after 36 iterations and we obtain the pseudo-partition defined in Table 5.7. The three cluster centres are $V_1 = (3.00, 3.00)$, $V_2 = (6.5, 7)$, and $V_3 = (11.5, 11.9)$, and $J_{\text{PCM}-\sigma} = 7.7836e-13$.

5.6 Problem set

In this study, we analyse the **physical growth patterns** of five students using their **height (cm)** and **weight (kg)**. The data contains a potential **noisy or outlier entry**. Using **PCM-σ**, we aim to group students into:

- **Healthy growth** (normal height and weight)
- **Underweight**
- **Overweight/obese**

The robustness of PCM-σ allows better handling of measurement errors or abnormal entries.

Table 1: Student physical profiles.

Student	Height (cm)	Weight (kg)
A	145	40
B	150	42
C	165	75
D	160	80
E	180	40 (noisy)

Research inquiries:

(i) **Noise tolerance:** How does PCM-σ assign the outlier student (E) compared to traditional PCM?

(ii) **Cluster comparison:** What does each cluster centre represent, and how does robust PCM adjust them?

(iii) **Robust typicality:** Which students are robustly associated with their clusters (high typicality), and which are not?

(iv) **Medical screening use:** How can Robust PCM-σ be useful in real-world health screening applications?

Problem 2

This study evaluates five students based on their scores in **science**, **math**, and **English**. One student shows inconsistent results, possibly due to recording errors or irregular learning patterns. Robust PCM is applied to minimize the effects of such noise.

Table 2: Students' academic performance.

Student	Science	Math	English
A	85	88	84
B	25	30	90
C	82	85	86
D	90	92	88
E	10	95	15 (noisy/misfit)

Research inquiries:

(i) **Role of robust PCM-σ:** How does PCM-σ handle student E, who has strong math but poor overall performance?

(ii) **Cluster centres analysis:** What kind of learners do the resulting cluster centres represent (e.g. strong STEM, balanced, weak)?

(iii) **Outlier detection:** How do typicality scores help identify whether student E is part of any valid cluster?

(iv) **Improved clustering accuracy:** Compare PCM vs. PCM-σ. How does robustification change the clustering results?

Problem 3

This example examines the **shopping behaviour** of five customers using four features: **Items bought**, **total spent**, **returns made**, and **loyalty points**. The dataset includes irregular behaviour, which may be due to fraud, special offers, or errors. Robust PCM helps provide more accurate clustering even with such inconsistencies.

Table 3: Customer shopping behaviour.

Customer	Items bought	Total spent (₹)	Returns	Loyalty points
Cust_1	10	1,000	1	200
Cust_2	2	500	0	450
Cust_3	12	1,100	2	180
Cust_4	1	200	10	0 (noisy)
Cust_5	11	1,050	1	190

Research inquiries:

(i) **Robustness to outlier behaviour:** Does PCM-σ treat customer 4 as a core or borderline member? How does the typicality value reflect this?

(ii) **Profile detection:** What kinds of buyer profiles are found (e.g. regular buyer, high-return customer, loyal shopper)?
(iii) **Improved cluster quality:** How does the inclusion of σ in robust PCM prevent misleading cluster assignments?
(iv) **Retail use case:** Why is robust clustering useful in analysing uncertain or noisy shopping data?

Chapter 6
Fuzzy possibilistic *C*-means (FPCM)

6.1 Introduction

The fuzzy possibilistic *C*-means (FPCM) algorithm intends to merge the fuzzy *C*-means (FCM) and possibilistic *C*-means (PCM) algorithms into one code. Although FCM defines membership values that can be used to define how strongly each data point is associated with each cluster, which is of assistance when working with overlapping data, it is rather prone to noise and outliers. Typicality values used in PCM, alternatively, demonstrate the degree to which a point represents an entire cluster, and has proved more robust to noise, although introducing their own kinds of problem, such as coincident clusters where two or more cluster centres collide.

This issue is solved by FPCM, which, during clustering, calculates both the membership and the typicality values of each data point simultaneously. Combining both information, FPCM gives the flexibility of fuzzy membership and, in addition, the noise resistance of possibilistic typicality. This translates to more realistic and stable clusters, and thus the algorithm is very practical and of interest in practical applications including medical image analysis, document clustering, and, more generally, whenever outliers and overlapping groups frequently occur in data.

In 1997, M. Keller and C. Bezdek proposed a groundbreaking approach called the FPCM method. The application of FPCM to unlabelled data results in the generation of membership and typicality values [24, 49, 50]. The constraint on row sums leads to imprecise results when assessing typicality in datasets of considerable size. The proposed optimization issue for the FPCM algorithm is as follows [51, 52]:

$$J_{\text{FPCM}} = \sum_{i=1}^{C}\sum_{k=1}^{N}(\mu_{ik}^{m} + t_{ik}^{\eta})\,\|x_k - v_i\|^2$$

$$\eta > 1 \text{ is possibilistic exponent (typically } \eta = 2)$$

and

$$\sum_{i=1}^{C}\mu_{ik} = 1\ \forall k \ \text{ and } \sum_{k=1}^{N} t_{ik} = 1\ \forall i$$

Following are the cluster centres, typicality grades and membership grades respectively that minimize the objective function:

$$\mu_{ik} = \left[\sum_{j=1}^{C}\left(\frac{D_{ik}}{D_{jk}}\right)^{2/m-1}\right]^{-1};\quad 1 \le i \le C;\ 1 \le k \le N$$

https://doi.org/10.1515/9783111494272-006

$$t_{ik} = \left[\sum_{j=1}^{N} \left(\frac{D_{ik}}{D_{ij}} \right)^{2/\eta-1} \right]^{-1}; \quad 1 \le i \le C; \; 1 \le k \le N$$

$$v_i = \frac{\sum_{k=1}^{N} (\mu_{ik}^m + t_{ik}^\eta) x_k}{\sum_{k=1}^{N} (\mu_{ik}^m + t_{ik}^\eta)}; \quad 1 \le i \le C$$

Here,

$$D_{ik}{}^2 = \|x_k - v_i\|^2$$

When the row-sum constraint is applied to large datasets, it leads to undesired levels of typicality. In contrast to PCM, FPCM does not consistently suffer from issues related to responsiveness.

6.2 Mathematical derivation for FPCM

Derivation for possibilistic membership t_{ik}

For finding possibilistic membership, we will use the Lagrangian multiplier method to incorporate the constraint:

$$\sum_{k=1}^{N} t_{ik} = 1$$

The Lagrangian for possibilistic membership is

$$L = \sum_{i=1}^{C} \sum_{k=1}^{N} (\mu_{ik}^m + t_{ik}^\eta) \|x_k - v_i\|^2 + \sum_{i=1}^{C} \lambda \left(\sum_{k=1}^{N} t_{ik} - 1 \right)$$

Now, compute the partial derivative of L with respect to t_{ik}:

$$\frac{\partial L}{\partial t_{ik}} = \eta t_{ik}^{\eta-1} \|x_k - v_i\|^2 + \lambda$$

Set it to zero:

$$\Rightarrow \eta t_{ik}^{\eta-1} \|x_k - v_i\|^2 = -\lambda$$

$$\Rightarrow t_{ik}^{\eta-1} = \frac{-\lambda}{\eta \|x_k - v_i\|^2}$$

$$=> t_{ik} = \left(\frac{-\lambda}{\eta\|x_k - v_i\|^2}\right)^{\frac{1}{\eta-1}} \tag{6.1}$$

To eliminate λ, we use the normalization constraint:

$$\sum_{k=1}^{N} t_{ik} \quad = 1$$

$$=> \sum_{k=1}^{N} t_{ik} = \sum_{k=1}^{N}\left(\frac{-\lambda}{\eta\|x_k - v_i\|^2}\right)^{\frac{1}{\eta-1}} = 1$$

$$=> \left(\frac{-\lambda}{\eta}\right)^{\frac{1}{\eta-1}} = \left(\sum_{k=1}^{N}\left(\frac{1}{\|x_k - v_i\|^2}\right)^{\frac{1}{\eta-1}}\right)^{-1}$$

Putting this value in (6.1), we get

$$=> t_{ik} = \left(\frac{1}{\|x_k - v_i\|^2}\right)^{\frac{1}{\eta-1}}\left(\sum_{j=1}^{N}\left(\frac{1}{\|x_j - v_i\|^2}\right)^{\frac{1}{\eta-1}}\right)^{-1}$$

$$=> t_{ik} = \left[\sum_{j=1}^{N}\left(\frac{D_{ik}}{D_{ij}}\right)^{2/\eta-1}\right]^{-1}$$

Derivation for membership μ_{ik}

For finding membership, we will again use the Lagrangian multiplier method to incorporate the constraint:

$$\sum_{i=1}^{C} \mu_{ik} \quad = 1$$

The Lagrangian for membership is

$$L' = \sum_{i=1}^{C}\sum_{k=1}^{N}(\mu_{ik}^{m} + t_{ik}^{\eta})\,\|x_k - v_i\|^2 + \sum_{k=1}^{N}\lambda'\left(\sum_{i=1}^{C}\mu_{ik} - 1\right)$$

Now, compute the partial derivative of L with respect to μ_{ik}:

$$\frac{\partial L}{\partial \mu_{ik}} = m\mu_{ik}^{m-1}\|x_k - v_i\|^2 + \lambda$$

Set it to zero:

$$=> m\mu_{ik}^{m-1}||x_k - v_i||^2 = -\lambda$$

$$=> \mu_{ik}^{m-1} = \frac{-\lambda}{m||x_k - v_i||^2}$$

$$=> \mu_{ik} = \left(\frac{-\lambda}{m||x_k - v_i||^2}\right)^{\frac{1}{m-1}} \tag{6.2}$$

To eliminate λ, we use a normalization constraint:

$$\sum_{i=1}^{C} \mu_{ik} = 1$$

$$=> \sum_{i=1}^{C} \mu_{ik} = \sum_{i=1}^{C} \left(\frac{-\lambda}{m||x_k - v_i||^2}\right)^{\frac{1}{m-1}} = 1$$

$$=> \left(\frac{-\lambda}{m}\right)^{\frac{1}{m-1}} = \left(\sum_{i=1}^{C} \left(\frac{1}{||x_k - v_i||^2}\right)^{\frac{1}{m-1}}\right)^{-1}$$

Putting this value in (6.2), we get

$$=> \mu_{ik} = \left(\frac{1}{||x_k - v_i||^2}\right)^{\frac{1}{m-1}} \left(\sum_{j=1}^{C} \left(\frac{1}{||x_k - v_j||^2}\right)^{\frac{1}{m-1}}\right)^{-1}$$

$$=> \mu_{ik} = \left[\sum_{j=1}^{C} \left(\frac{D_{ik}}{D_{jk}}\right)^{2/m-1}\right]^{-1}$$

Derivation for cluster centre v_i

Now, for finding the cluster centre, we take the derivative of J_{FPCM} with respect to v_i and then set it to zero:

$$\frac{\partial J_{\text{FPCM}}}{\partial v_i} = \sum_{k=1}^{N} -2\ (\mu_{ik}^m + t_{ik}^\eta)(x_k - v_i) = 0$$

$$=> \sum_{k=1}^{N} (\mu_{ik}^m + t_{ik}^\eta) x_k = \sum_{k=1}^{N} (\mu_{ik}^m + t_{ik}^\eta) v_i$$

$$=> v_i = \frac{\sum_{k=1}^{N}(\mu_{ik}^m + t_{ik}^\eta)x_k}{\sum_{k=1}^{N}(\mu_{ik}^m + t_{ik}^\eta)}$$

FPCM:

---

m=2; %fuzziness index
n=2; %possibilistic exponent

Take initial cluster centres as the centres obtained from FCM algorithm;

Start "For" loop for number of iterations (say r);
Take data set;
Find distance 'd' between cluster centres and data points using;

Again start "For" loop for finding membership values iteratively as many times as number of data points.
Find membership values using;

$$\mu_{ik} = \left[\sum_{j=1}^{C}\left(\frac{D_{ik}}{D_{jk}}\right)^{2/m-1}\right]^{-1}$$

End the loop for membership values;

Again start "For" loop for finding typicality values iteratively as many times as number of data points.
Find typicality values using;

$$t_{ik} = \left[\sum_{j=1}^{N}\left(\frac{D_{ik}}{D_{ij}}\right)^{2/\eta-1}\right]^{-1}$$

End the loop for typicality values;
In next step find out values of clusters centres using;

$$V_i = \frac{\sum_{k=1}^{N}(\mu_{ik}^m + t_{ik}^\eta)x_k}{\sum_{k=1}^{N}(\mu_{ik}^m + t_{ik}^\eta)}$$

In next step put the terminating condition for checking optimization of solution.

End the "For" loop for number of iterations 'r';

In next step find the value of objective function;

$$J_{FPCM} = \sum_{i=1}^{C}\sum_{k=1}^{N}(\mu_{ik}^{m} + t_{ik}^{\eta})\|x_k - v_i\|^2$$

And plot the data points and centres on the graph for better analysis of data

6.3 Python code

```
import numpy as np
import matplotlib.pyplot as plt

def fpcm(X, C=3, m=2, eta=2, max_iter=100, tol=1e-5, random_state=42):
    np.random.seed(random_state)
    N = X.shape[0]

    # Step 1: Initialize membership from random (or could use FCM first)
    mu = np.random.rand(C, N)
    mu /= np.sum(mu, axis=0, keepdims=True)

    # Initialize typicality matrix randomly
    t = np.random.rand(C, N)
    t /= np.sum(t, axis=0, keepdims=True)

    obj_values = []

    for iteration in range(max_iter):
        # Step 2: Compute cluster centers
        numerator = np.dot((mu**m + t**eta), X)
        denominator = np.sum(mu**m + t**eta, axis=1, keepdims=True)
        v = numerator / (denominator + 1e-8)

        # Step 3: Compute distances
        D = np.zeros((C, N))
        for i in range(C):
            D[i] = np.linalg.norm(X - v[i], axis=1) + 1e-8 # avoid zero di-
vision

        # Step 4: Update membership values
        mu_new = np.zeros_like(mu)
        for k in range(N):
```

```
            for i in range(C):
                denom = np.sum( (D[i,k]/D[:,k]) ** (2/(m-1)) )
                mu_new[i,k] = 1 / (denom + 1e-8)

        # Step 5: Update typicality values
        t_new = np.zeros_like(t)
        for k in range(N):
            for i in range(C):
                denom = np.sum( (D[i,k]/D[i,:]) ** (2/(eta-1)) )

                t_new[i,k] = 1 / (denom + 1e-8)

        # Step 6: Compute objective function
        J = np.sum( (mu_new**m + t_new**eta) * (D**2) )
        obj_values.append(J)

        # Step 7: Check convergence
        if np.linalg.norm(mu_new - mu) < tol and np.linalg.norm(t_new - t)
< tol:
            print(f'Converged at iteration {iteration+1}')
            break

        mu = mu_new
        t = t_new

    # Plot objective over iterations
    plt.figure(figsize=(6,4))
    plt.plot(obj_values, '-o')
    plt.title('Objective Function over Iterations')
    plt.xlabel('Iteration')
    plt.ylabel('J_FPCM')
    plt.show()

    # Plot clustering result
    plt.figure(figsize=(6,6))
    cluster_labels = np.argmax(mu, axis=0)
    plt.scatter(X[:,0],    X[:,1],    c=cluster_labels,    cmap='viridis',
alpha=0.6)
    plt.scatter(v[:,0], v[:,1], c='red', s=200, marker='X', label='-
Centers')
    plt.title('FPCM Clustering Result')
    plt.legend()
    plt.show()
```

```
    return v, mu, t, obj_values

# Example usage:
if __name__ == "__main__":
    np.random.seed(0)
    X = np.vstack([
        np.random.randn(50,2) + [2,2],
        np.random.randn(50,2) + [-2,-2],
        np.random.randn(50,2) + [4,-3]
    ])
    centers, memberships, typicality, history = fpcm(X, C=3, max_iter=100)
    print("Final centers:\n", centers)
```

6.4 R-code

```
set.seed(42)

fpcm <- function(X, C=3, m=2, eta=2, max_iter=100, tol=1e-5) {
  N <- nrow(X)
  D <- ncol(X)
  mu <- matrix(runif(C*N), nrow=C)
  mu <- mu / colSums(mu)

  t <- matrix(runif(C*N), nrow=C)
  t <- t / colSums(t)

  obj_values <- numeric(max_iter)

  for (iter in 1:max_iter) {
    # Step 1: compute centers
    numerator <- (mu^m + t^eta) %*% X
    denominator <- rowSums(mu^m + t^eta)
    v <- numerator / (denominator + 1e-8)

    # Step 2: compute distances
    Dmat <- matrix(0, nrow=C, ncol=N)
    for (i in 1:C) {
      Dmat[i,] <- sqrt(rowSums((X - matrix(v[i,], nrow=N, ncol=D,
byrow=TRUE))^2)) + 1e-8
    }
```

```
    # Step 3: update membership
    mu_new <- matrix(0, nrow=C, ncol=N)
    for (k in 1:N) {
      for (i in 1:C) {
        denom <- sum( (Dmat[i,k] / Dmat[,k])^(2/(m-1)) )
        mu_new[i,k] <- 1 / (denom + 1e-8)
      }
    }

    # Step 4: update typicality
    t_new <- matrix(0, nrow=C, ncol=N)
    for (k in 1:N) {
      for (i in 1:C) {
        denom <- sum( (Dmat[i,k] / Dmat[i,])^(2/(eta-1)) )
        t_new[i,k] <- 1 / (denom + 1e-8)
      }
    }

    # Step 5: compute objective
    J <- sum( (mu_new^m + t_new^eta) * (Dmat^2) )
    obj_values[iter] <- J

    # Step 6: check convergence
    if (sqrt(sum((mu_new - mu)^2)) < tol && sqrt(sum((t_new - t)^2)) <
tol) {
      cat("Converged at iteration", iter, "\n")
      break
    }

    mu <- mu_new
    t <- t_new
  }

  # Plot objective
  plot(obj_values[1:iter], type="o", main="Objective Function over
Iterations",
       xlab="Iteration", ylab="J_FPCM")

  # Plot clustering
  cluster <- apply(mu, 2, which.max)
  plot(X, col=cluster, pch=19, main="FPCM Clustering Result")
  points(v, col=1:C, pch=8, cex=2)
```

```
  list(centers=v, membership=mu, typicality=t, objective=obj_values
[1:iter])
}
# Example usage
X <- rbind(
  matrix(rnorm(100, mean=2), ncol=2),
  matrix(rnorm(100, mean=-2), ncol=2),
  matrix(rnorm(100, mean=4), ncol=2)
)
result <- fpcm(X, C=3, max_iter=100)
print(result$centers)
```

6.5 Solved examples

Example 1:

The study investigates 15 CSK IPL players' stats including batting averages, over rates, and selection positions shown in table 6.1. Employing the FPCM clustering technique, we want to categorize the players into three unique clusters. Our analysis groups players into batters, all-rounders, and fast bowlers. Through this technique, we discover the strongest bowling all-rounder performers within our dataset.

Table 6.1: CSK team members' batting and bowling statistics along with their early designations.

Player	Batting average	Bowl average	Initial class
MM Ali	24.96	22.78	Batsman
S Anirudha	19	50	Bowler
R Ashwin	9.5	24.22	Bowler
S Badrinath	30.65	50	Batsman
L Balaji	3.66	26.09	Bowler
DE Bollinger	21	18.72	Batsman
DJ Bravo	25.1	22.47	Batsman
DL Chahar	10.83	28.2	Bowler
SM Curran	22	30.54	Batsman
MS Dhoni	40.53	50	Batsman
F du Plessis	35.33	50	Batsman
RD Gaikwad	39.36	50	Batsman
MS Gony	12.5	30.73	Bowler
Harbhajan Singh	7.5	25.3	Bowler

Research inquiries:

(i) **Classification of players:**

What is the use of FPCM clustering for categorizing players into three distinct categories based on their batting and bowling averages: Our system identifies players into three separate groups, which are batsmen, bowling all-rounders, and bowlers.

(ii) **Determination of optimal bowling all-rounders:**

What are the profiles of players in our sample who perform well both as bowlers and batsmen, and how do they stack up versus other player types?

(iii) **Cluster analysis:**

How does each performance segment appear different through its bowling and batting statistics?

The discovered clusters show how much they match the player categories from step one.

(iv) **Performance analysis:**

By using FPCM clustering, we can find player performance patterns that we cannot see through simple data review.

Solution:

Step 1: Consider batting average = X1 and bowling average = X2.

X1	24.96	19	9.5	30.65	3.66	21	25.1	10.83	22	40.53	35.33	39.36	12.5	7.5
X2	22.78	50	24.22	50	26.09	18.72	22.47	28.2	30.54	50	50	50	30.73	25.3

Step 2: The algorithm is initialized by randomly assigning membership values and typicality values across three cluster centres, forming the initial fuzzy and possibilistic partitions, respectively, as shown below.

T1	0.009	0.233	0.954	0.019	0.914	0.113	0.012	0.972	0.213	0.019	0.000	0.012	0.851	0.976
T2	0.003	0.566	0.006	0.957	0.017	0.020	0.004	0.004	0.070	0.953	0.996	0.969	0.025	0.004
T3	0.988	0.200	0.040	0.024	0.068	0.867	0.984	0.024	0.717	0.029	0.000	0.019	0.125	0.020

Step 3: Take $\eta = 2$.

T1^2	T2^2	T3^2	X1*T1^2	X2*T2^2	X1*T3^2	X2*T2^2	X1*T3^2	X2*T3^2
8.91E-05	9.32E-06	0.975172	0.002224	0.002029637	0.000233	0.000212	24.34029	22.21441982
0.05444	0.320807	0.040111	1.034363	2.722008372	6.095334	16.04035	0.762111	2.005554055
0.909592	3.58E-05	0.001623	8.641121	22.03031173	0.00034	0.000868	0.01542	0.039312236
0.00035	0.91563	0.000596	0.010713	0.01747694	28.06405	45.78148	0.018276	0.029813494
0.83604	0.000306	0.004645	3.059906	21.8122833	0.00112	0.007985	0.017	0.121184326
0.012696	0.000417	0.75152	0.266622	0.237674597	0.008756	0.007806	15.78191	14.06844884
0.000152	1.62E-05	0.96755	0.003817	0.003417342	0.000407	0.000364	24.28549	21.74083743

(continued)

T1^2	T2^2	T3^2	X1*T1^2	X2*T2^2	X1*T3^2	X2*T2^2	X1*T3^2	X2*T3^2
0.944111	1.74E-05	0.000585	10.22472	26.62392766	0.000188	0.00049	0.006331	0.01648642
0.045331	0.004912	0.514096	0.997284	1.384412102	0.108059	0.150005	11.31012	15.70050067
0.000353	0.907604	0.000814	0.014316	0.017660613	36.78521	45.38022	0.032974	0.040678988
0	0.991394	0	0	0	35.02595	49.56971	0	0
0.000154	0.938844	0.000348	0.006059	0.00769727	36.95291	46.94221	0.013694	0.017396339
0.723552	0.000611	0.01554	9.044404	22.23476191	0.007638	0.018778	0.194256	0.47755799
0.951905	1.5E-05	0.000419	7.139289	24.08320183	0.000112	0.000379	0.003144	0.010607031
4.478766	**4.080619**	**3.273019**	**40.44484**	**121.1768633**	**143.0503**	**203.9009**	**76.78103**	**76.48279764**

Step 4: Cluster centers

v1	9.188	26.86
v2	35.45	49.93
v3	23.69	23.06

Step 5: Typicality values

NT1	0.0840	0.03710	0.7748	0.02381	0.4385	0.1057	0.0820	0.8443	0.1204	0.0157	0.0195	0.0165	0.4841	0.8233
NT2	0.04096	0.11793	0.02640	0.61065	0.0224	0.02968	0.04032	0.03246	0.06101	0.58409	0.99945	0.70338	0.03884	0.02543
NT3	0.94148	0.03527	0.11885	0.03410	0.06246	0.51166	0.92094	0.12478	0.31751	0.02636	0.03076	0.02737	0.12937	0.09305

Step 6: $e = 0.0001$.

e1	1.2345e-05	1.2802e-05	1.0908e-05	2.4357e-06	1.5862e-05	1.2342e-05	1.1061e-05	1.4344e-05	1.3685e-05	1.5236e-06	2.0403e-09	1.1061e-05	2.4724e-05	1.5305e-05
e2	1.2825e-05	1.2845e-05	1.2797e-05	2.3711e-05	1.3567e-05	1.2406e-05	1.3758e-05	1.6759e-05	1.2974e-05	1.3806e-05	1.5774e-05	1.5746e-05	1.2935e-05	1.2759e-05
e3	3.7487e-06	1.3903e-05	1.2697e-05	1.4705e-05	1.2668e-05	2.7152e-06	3.8749e-06	1.4706e-05	1.3326e–05	1.6706e-05	3.0403e-09	1.0961e-05	1.2945e-05	1.2759e-05

Example 2

An organization has warehouses for life-saving drugs at six different locations, whose coordinates are given in the following table. The organization is in the process of locating one more critical central warehouse, which will distribute drugs to all the existing warehouses on emergency request. Find the number of optimum locations of the new facility (warehouse) based on the FPCM clustering concept.

Table 6.2: Thirty warehouses whose coordinates are summarized.

Existing warehouse number	Coordinates of centroids
1	(2.5,3.5)
2	(2.7,3.6)
3	(2.2,3.2)
4	(3.0,3.0)
5	(2.7,3.1)
6	(2.3,3.5)
7	(3.1,3.0)
8	(3.3,2.9)
9	(3.0,3.5)
10	(2.9, 2.9)
11	(6.0,7.2)
12	(6.5,7.0)
13	(5.9,6.9)
14	(6.3,7.3)
15	(5.9,7.0)
16	(6.8,7.0)
17	(6.5,6.9)
18	(6.2,7.1)
19	(7.2,7.2)
20	(6.6,7.1)
21	(11.1,13.5)
22	(12.2,12.5)
23	(11.5,11.9)

Solution:

The data are shown in Table 6.2. Assume that we want to determine a fuzzy pseudo-partition with three clusters (i.e. $c = 3$). Assume further that we choose $m = 2$ and $||.||$ is the Euclidean distance. The algorithm is initialized by randomly assigning membership values and typicality values across three cluster centres, forming the initial fuzzy and possibilistic partitions, respectively, is $T^{(0)} = [T_1, T_2, T_3]$. Table 6.3–6.7 shows the results obtained at different iterations.

Table 6.3: Cluster centres and the corresponding typicality function values of matrix *T*.

Input data		**Typicality function for fuzzy clustering**		
X_1	X_2	T_1	T_2	T_3
X_1	X_2	T_1	T_2	T_3
2.5	3.5	.9	.05	.05
2.7	3.6	.89	.055	.055
2.2	3.2	.88	.06	.06
3.0	3.0	.87	.065	.065
2.7	3.1	.86	.07	.07
2.3	3.5	.84	.09	.07
3.1	3.0	.81	.11	.08
3.3	2.9	.8	.1	.1
3.0	3.5	.79	.11	.1
2.9	2.9	.75	.15	.1
6.0	7.2	.05	.9	.05
6.5	7.0	.055	.89	.055
5.9	6.9	.06	.88	.06
6.3	7.3	.065	.87	.065
5.9	7.0	.07	.86	.07
6.8	7.0	.09	.84	.07
6.5	6.9	.11	.81	.08
6.2	7.1	.1	.8	.1
7.2	7.2	.11	.79	.1
6.6	7.1	.15	.75	.1
11.1	13.5	.05	.05	.9
12.2	12.5	.055	.055	.89
11.5	11.9	.06	.06	.88
13.2	13.3	.065	.065	.87
11.5	12.5	.07	.07	.86
12.0	15.0	.07	.09	.84
13.2	13.2	.08	.11	.81
11.2	11.9	.1	.1	.8
12.9	13.7	.1	.11	.79
13.9	13.0	.1	.15	.75

Iteration 1:

Table 6.4: Cluster centres V and the corresponding typicality function values of matrix T (Iteration 1).

Cluster centres		V_1	V_2	V_3
		(4.129, 4.688)	**(7.339, 8.143)**	**(10.13, 10.83)**
Average values		**(4.1, 4.7)**	**(7.3, 8.1)**	**(10.1, 10.8)**
Data points		T_1	T_2	T_3
2.5	3.5	0.907354932617891	0.426504032170627	0.330913592375793
2.7	3.6	0.925046136794869	0.442379599181050	0.339995040184233
2.2	3.2	0.870265750243584	0.396805620896120	0.313743469534404
3.0	3.0	0.906143602274280	0.424858000132173	0.330534107302721
2.7	3.1	0.897161068818946	0.416007263879364	0.325070557143724
2.3	3.5	0.893275469022726	0.416023076889515	0.324905826661513
3.1	3.0	0.910616578719732	0.429538520929033	0.333353816859759
3.3	2.9	0.911119235479149	0.432969634213109	0.335776636708569
3.0	3.5	0.936797401984510	0.453014348725417	0.346192758383486
2.9	2.9	0.894270482634313	0.414761253566208	0.324695285889544
6.0	7.2	0.802340083119272	0.925760225206171	0.646800230992749
6.5	7.0	0.784078821336986	0.943303834678950	0.665406256826573
5.9	6.9	0.832210434421432	0.902441263891274	0.624178236280245
6.3	7.3	0.775392332853080	0.949196265461412	0.671149471029375
5.9	7.0	0.824415628240771	0.908288196479959	0.629681234566177
6.8	7.0	0.761403987882326	0.954433107551154	0.682553529474434
6.5	6.9	0.791126262735759	0.936998964017487	0.659264164299113
6.2	7.1	0.797585205548790	0.933435029710561	0.653506182521370
7.2	7.2	0.716709348618569	0.973558531136247	0.717933928416288
6.6	7.1	0.769583805620377	0.953425072724523	0.677389733320361
11.1	13.5	0.239790148887811	0.438410924403500	0.872542498203069
12.2	12.5	0.239903589511981	0.439737667463119	0.886693455146537
11.5	11.9	0.272441053525325	0.515543954787953	0.948241359461573
13.2	13.3	0.202890201738732	0.354330157428809	0.780968544056034
11.5	12.5	0.256614432489779	0.479555919438287	0.922188837333689
12.0	15.0	0.191345771423889	0.327291909362093	0.725844094630766
13.2	13.2	0.204675992827996	0.358206317621418	0.786347449939571
11.2	11.9	0.280763274761431	0.535404881146792	0.960246640335751
12.9	13.7	0.201150746792437	0.351134241322777	0.776699199541020
13.9	13.0	0.194837449904686	0.334181899192816	0.745340338874829

Iteration 2:

Table 6.5: Cluster centres V and the corresponding typicality function values of matrix T (Iteration 2).

Cluster centres		V_1	V_2	V_3
		(4.27, 4.813)	(6.438, 7.076)	(10.43, 11.16)
Average values		(4.3, 4.8)	(6.4, 7.1)	(10.4, 11.2)
Data points		T_1	T_2	T_3
2.5	3.5	0.538500853455296	0.0735856749268610	0.0613399274849299
2.7	3.6	0.590120333825608	0.0794102753877903	0.0636255315822291
2.2	3.2	0.451436627148977	0.0637935131913978	0.0571352476277453
3.0	3.0	0.536321607641860	0.0732529780422327	0.0612305377802624
2.7	3.1	0.512115180286107	0.0701716333354432	0.0598807586461766
2.3	3.5	0.502766689048547	0.0698901167094645	0.0598560629577735
3.1	3.0	0.548999201269946	0.0749070748287245	0.0619348527880886
3.3	2.9	0.551965488157939	0.0760985576006207	0.0625440027950271
3.0	3.5	0.629408579861472	0.0836898021968763	0.0652050024881114
2.9	2.9	0.505848513635919	0.0697892702736145	0.0597850122709113
6.0	7.2	0.394740706939146	0.915715256039925	0.183578051977837
6.5	7.0	0.367482900252908	0.995692575431651	0.195096547396637
5.9	6.9	0.446985146790411	0.875266848359718	0.170342788551758
6.3	7.3	0.354822630237063	0.970240996168804	0.198980441696467
5.9	7.0	0.432409558793717	0.883973808317332	0.173475374116644
6.8	7.0	0.336343671578274	0.942518425977731	0.206602548811028
6.5	6.9	0.377957013121295	0.984659464145413	0.191143335432596
6.2	7.1	0.387599962096792	0.975235398399366	0.187607649997722
7.2	7.2	0.284101108116852	0.790327697784887	0.233080930516042
6.6	7.1	0.347149332267200	0.988181707117145	0.203066025950697
11.1	13.5	0.0443556587632133	0.0344487566110770	0.573805961703236
12.2	12.5	0.0444037881738237	0.0346506168930693	0.618126066312336
11.5	11.9	0.0523991783393363	0.0439503517441317	0.825317603830334
13.2	13.3	0.0359993938892864	0.0259219695286205	0.394038229355127
11.5	12.5	0.0484301563075019	0.0392335957710235	0.730945088701594
12.0	15.0	0.0334984106253943	0.0234206047982599	0.316397720726878
13.2	13.2	0.0363894091310292	0.0262964254204422	0.402360135503626
11.2	11.9	0.0545411459918777	0.0466382265591161	0.875367337574165
12.9	13.7	0.0356189171406569	0.0255764190833318	0.388296006438925
13.9	13.0	0.0342607790133297	0.0241627301033519	0.340484554425642

Due to space and readability considerations, it is impractical to present the complete set of iterative results. Therefore, the following tables provide a summary of the membership values and cluster centres obtained during the penultimate (i.e. 15th) and final (i.e. 16th) iterations, which effectively illustrate the convergence behaviour of the algorithm.

Iteration 15:

Table 6.6: Cluster centres V and the corresponding typicality function values of matrix T (Iteration 15).

Cluster centres		V_1	V_2	V_3
		(2.779, 3.214)	**(6.388 7.071)**	**(12.31 13.02)**
Average values		**(2.8, 3.2)**	**(6.4, 7.1)**	**(12.3, 13.02)**
Data points		T_1	T_2	T_3
2.5	3.5	0.560002747341550	0.00783453123909448	0.00848082317823244
2.7	3.6	0.566812623950015	0.00850661450668535	0.00874904238129197
2.2	3.2	0.377240788276385	0.00672074792721101	0.00798195423204027
3.0	3.0	0.682103239518592	0.00778386694979541	0.00847106401068649
2.7	3.1	0.913571496238471	0.00743701391822720	0.00830903112636377
2.3	3.5	0.394967916424638	0.00741353968763396	0.00830616662673720
3.1	3.0	0.577042890816301	0.00797210760637000	0.00855500982077954
3.3	2.9	0.354345262713921	0.00810452933169243	0.00863021087152337
3.0	3.5	0.608495267839674	0.00900012827082200	0.00893386806314009
2.9	2.9	0.642051806956736	0.00738879388667655	0.00830010510934143
6.0	7.2	0.00767475417123322	0.567945446616833	0.0212324575584629
6.5	7.0	0.00715649505661800	0.926227428502429	0.0223260069408614
5.9	6.9	0.00863228589598991	0.451034460762873	0.0199458993604108
6.3	7.3	0.00693342832798079	0.785427913687145	0.0226942284380043
5.9	7.0	0.00836661335159425	0.474626297429481	0.0202529163371803
6.8	7.0	0.00661517031237918	0.557814788575366	0.0234371146380705
6.5	6.9	0.00734998481293787	0.840439832882949	0.0219534986053254
6.2	7.1	0.00752095532852519	0.858334322815346	0.0216078890493823
7.2	7.2	0.00569980831804431	0.245834410746711	0.0259545865935135
6.6	7.1	0.00679240755769024	0.828410634580154	0.0230824354994563
11.1	13.5	0.00115909841617776	0.00345273478575645	0.487277525969595
12.2	12.5	0.00115946655177738	0.00346807452708217	0.848847984730023
11.5	11.9	0.00133893991451619	0.00443176922656160	0.455534925523477
13.2	13.3	0.000964837243973046	0.00257687903789367	0.645811579766457
11.5	12.5	0.00125008508787105	0.00394313634169644	0.633978319482281
12.0	15.0	0.000906243929335205	0.00232729974595447	0.285339226388483
13.2	13.2	0.000974124593656551	0.00261470754631863	0.657889256465168
11.2	11.9	0.00138592050325336	0.00471435080192239	0.391533100098957
12.9	13.7	0.000955460268564510	0.00254276222547492	0.663184924005101
13.9	13.0	0.000924785426810603	0.00239780494645111	0.385995192831776

Iteration 16:

Table 6.7: Cluster centres V and the corresponding typicality function values of matrix T (Iteration 16).

Cluster centres		V_1	V_2	V_3
		(2.779, 3.214)	(6.388, 7.071)	(12.31, 13.02)
Average values		(2.78, 3.21)	(6.4, 7.1)	(12.3, 13.0)
Data points		T_1	T_2	T_3
2.5	3.5	0.560	0.00783	0.00848
2.7	3.6	0.5668	0.00850	0.00875
2.2	3.2	0.3772	0.00672	0.00798
3.0	3.0	0.6821	0.007784	0.00847
2.7	3.1	0.9135	0.007437	0.00831
2.3	3.5	0.39497	0.007414	0.00831
3.1	3.0	0.5770	0.007972	0.00855
3.3	2.9	0.3543	0.008105	0.00863
3.0	3.5	0.6085	0.00900	0.00893
2.9	2.9	0.6420	0.0073895	0.00829
6.0	7.2	0.00767	0.56796	0.02123
6.5	7.0	0.00715	0.92623	0.022324
5.9	6.9	0.0086	0.45105	0.019944
6.3	7.3	0.0069	0.785445	0.022692
5.9	7.0	0.00836	0.47464	0.020251
6.8	7.0	0.0066	0.55784	0.023435
6.5	6.9	0.00735	0.84045	0.021951
6.2	7.1	0.00752	0.85834	0.021606
7.2	7.2	0.00569	0.24585	0.025952
6.6	7.1	0.00679	0.82843	0.023080
11.1	13.5	0.00115	0.00345	0.487197
12.2	12.5	0.00115	0.00346	0.848728
11.5	11.9	0.00133	0.00443	0.455416
13.2	13.3	0.000964	0.00257	0.645978
11.5	12.5	0.00125	0.00394	0.633806
12.0	15.0	0.0009	0.00232	0.285368
13.2	13.2	0.00097	0.00261	0.658052
11.2	11.9	0.00138	0.00471	0.391428
12.9	13.7	0.00095	0.00254	0.663353
13.9	13.0	0.00092	0.00239	0.3860825

Then, the algorithm stops after 16 iterations, and we obtain the pseudo-partition defined in Table 6.7. The three cluster centres are $V_1 = (2.779, 3.214)$, $V_2 = (6.388, 7.071)$, and $V_3 = (12.31, 13.02)$, and $J_{\text{FPCM}} = 24.0782$.

6.6 Problem set

Problem 1

We assess the **physical activity** of five students using two features: **Walking hours per week** and **screen time per day**. The **FPCM** algorithm is used to classify students into:
- **Active lifestyle** (high walking, low screen time)
- **Sedentary lifestyle** (low walking, high screen time)
- **Balanced lifestyle** (moderate in both)

Unlike FCM or PCM alone, FPCM uses **both membership and typicality** for better clustering.

Table 1: Physical activity data.

Student	Walking hours/week	Screen time (h/day)
A	6.0	2.0
B	1.0	8.5
C	4.0	3.5
D	7.5	1.5
E	2.0	7.0

Research inquiries:
(i) **Hybrid participation detection:** Which students have fuzzy memberships in multiple clusters but high typicality in one?
(ii) **Interpretation of role:** What do the cluster centres tell us about healthy, moderate, and sedentary behaviour?
(iii) **Value of dual measures:** Why is combining membership and typicality more insightful than using only one?
(iv) **Improved lifestyle insights:** How does FPCM reveal hidden patterns in behaviour compared to FCM or PCM?

Problem 2

This study profiles five students based on three aspects: **Reading time**, **visual learning time**, and **group study time**. FPCM clusters them into:
- **Visual learners**
- **Independent learners**
- **Collaborative learners**

The strength of FPCM lies in capturing **overlapping learning habits** and the **trust level (typicality)** of each student's cluster assignment.

Table 2: Student learning style data.

Student	Reading (h)	Visual (h)	Group study (h)
A	5	2	1
B	2	6	1.5
C	3	2	5
D	4	3	2
E	1	5	4

Research inquiries:
(i) **Overlap between styles:** Which students demonstrate fuzzy participation in more than one learning group?h
(ii) **Typicality score meaning:** How does high typicality help in confirming a student's learning preference?
(iii) **Cluster profile summary:** What does each cluster centre represent in terms of learning behaviour?
(iv) **Practical use in education:** How can schools use FPCM to assign students to study groups?

Problem 3

Five employees are assessed on **task completion**, **team contribution**, **independent work**, and **stress level**. FPCM is used to group them into:
- **Team players**
- **Independent contributors**
- **High-stress workers**

Using both **fuzzy membership** and **typicality**, FPCM gives a fuller picture of each employee's style, including ambiguous or overlapping roles.

Table 3: Employee's work behaviour.

Employee	Task completion (%)	Team contribution	Independent work	Stress level
E1	95	90	60	30
E2	70	85	40	35
E3	85	20	90	60
E4	60	25	80	85
E5	80	75	70	40

Research inquiries:

(i) **Cluster blending:** Which employees fit into multiple profiles with varying degrees?

(ii) **Reliability via typicality:** Who has high fuzzy membership but low typicality – and what does it mean?

(iii) **Understanding work patterns:** What do the cluster centres reveal about typical employee roles?

(iv) **Organizational application:** How can FPCM help HR teams in employee placement or support planning?

Chapter 7
Possibilistic fuzzy *C*-means (PFCM)

7.1 Introduction

One such algorithm is possibilistic fuzzy *C*-means (PFCM), a combination of fuzzy *C*-means (FCM) and possibilistic *C*-means (PCM) algorithms with their stronger characteristics. Data points are allowed to be partially represented in any number of clusters depending upon membership values in FCM, which is useful to cluster overlapping data but leaves the algorithm susceptible to noise and outliers. PCM, however, has typicality values, which describe the extent to which a point represents a cluster; hence, is more resistant to noise as this can cause coincident clusters (when two or more clusters converge to a common centre).

PFCM overcomes these potential issues by calculating both membership and typicality values at once in each iteration. With this dual representation, the algorithm is able to maintain the flexibility of fuzzy clustering, enhance its tolerance to outliers, and minimize the likelihood of coincident clusters. Accordingly, PFCM generates more robust and explainable groupings, which is also the reason PFCM finds great use in practical cases like image segmentation, bioinformatics, and pattern recognition, where data is usually contaminated or overlapping.

In addition to the conventional point prototypes for each cluster, PFCM concurrently creates memberships and possibilities [50, 52, 53]. PFCM, which combines elements of PCM and FCM, is often employed as a solution to address the various challenges encountered in PCM, FCM, and FPCM [54]. PFCM effectively addresses the coincident clusters problem encountered in PCM, the noise sensitivity limitation observed in FCM, and the row sum constraints inherent in FPCM. Typicality values are constrained by FPCM, which is an obvious problem. For convenience, we relax the restriction on membership values (column sum = 1) while leaving the row sum = 1 restriction in place. Since this is the case, we are left with the optimization problem shown below:

$$J_{\text{PFCM}} = \sum_{k=1}^{N}\sum_{i=1}^{C}\left(a\mu_{ik}^{m} + bt_{ik}^{\eta}\right) \times \|x_k - v_i\|^2 + \sum_{i=1}^{C} Y_i \sum_{k=1}^{N} (1 - t_{ik})^{\eta}$$

Subject to the constraints $m > 1, \eta > 1, 0 \le \mu_{ik}, t_{ik} \le 1, \sum_{i=1}^{C} \mu_{ik} = 1 \ \forall \ k$.

The constants a and b in the method represent user-defined values that indicate the relative significance of the fuzzy and possibilistic elements. It is common practice to assign both a and b a value of 1. The algorithm's accuracy is enhanced by larger values of b.

https://doi.org/10.1515/9783111494272-007

Following are the cluster centres, typicality grades and membership grades, respectively, that minimize the objective function:

$$\mu_{ik} = \left[\sum_{j=1}^{C}\left(\frac{D_{ik}}{D_{jk}}\right)^{2/m-1}\right]^{-1}; \quad 1 \le i \le C; 1 \le k \le N$$

$$t_{ik} = \frac{1}{1+\left(\frac{b\,D_{ikA}^2}{Y_i}\right)^{\frac{1}{\eta-1}}}; \quad 1 \le i \le C; 1 \le k \le N$$

$$v_i = \frac{\sum_{k=1}^{N}\left(a\mu_{ik}^m + bt_{ik}^\eta\right)x_k}{\sum_{k=1}^{N}\left(a\mu_{ik}^m + bt_{ik}^\eta\right)}; \quad 1 \le i \le C$$

$$Y_i = \frac{\sum_{j=1}^{N} t_{ij}^\eta \left\|x_j - v_i\right\|^2}{\sum_{j=1}^{N}\left(1-t_{ij}\right)^\eta}$$

The PFCM algorithm commences by initializing cluster centres, which are determined by the FCM algorithm. Subsequently, it proceeds to update the cluster centres v_i, membership degrees μ_{ij}, and typicalities t_{ij} until convergence conditions are satisfied. It should be noted that when the PFCM method is implemented with the values of $a = 1$ and $b = 0$, it effectively transforms into the FCM algorithm. The data points that are located significantly distant from the cluster centres exhibit reduced typicalities and hence exert a diminished influence on the cluster centres. This approach is deemed more suitable for managing noise and outliers due to the following reasons. Nevertheless, this technique has a susceptibility to initialization and the presence of contemporaneous clusters; particularly in scenarios where the input contains substantial noise [55, 56] and the clusters overlap.

7.2 Mathematical derivation for robust PFCM

Derivation for membership μ_{ik}

The typical robust PFCM objective function is given above as J_{PFCM}. But we're now focusing only on the fuzzy membership μ_{ik} part of this function:

$$J_\mu = \sum_{k=1}^{N}\sum_{i=1}^{C}\left(a\mu_{ik}^m\right) \times \left\|x_k - v_i\right\|^2$$

Subject to

$$\sum_{i=1}^{c} \mu_{ik} = 1$$

To impose the constraint on fuzzy memberships, we construct the Lagrangian:

$$L = \sum_{k=1}^{N} \sum_{i=1}^{C} (a\mu_{ik}^{m}) \times \|x_k - v_i\|^2 - \lambda\left(\sum_{i=1}^{c} \mu_{ik} - 1\right)$$

For finding μ_{ik}, compute the partial derivative of L with respect to μ_{ik}:

$$\frac{\partial L}{\partial \mu_{ik}} = am\mu_{ik}^{m-1}\|x_k - v_i\|^2 - \lambda$$

Set it to zero:

$$\Rightarrow am\mu_{ik}^{m-1}\|x_k - v_i\|^2 - \lambda = 0$$

$$\Rightarrow \mu_{ik}^{m-1} = \frac{\lambda}{am\|x_k - v_i\|^2}$$

$$\Rightarrow \mu_{ik} = \left(\frac{\lambda}{am\|x_k - v_i\|^2}\right)^{\frac{1}{m-1}}$$

Now for finding the value of λ, enforce the constraint:

$$\sum_{i=1}^{c} \mu_{ik} = 1$$

$$\Rightarrow \sum_{i=1}^{c} \mu_{ik} = \sum_{i=1}^{c} \left(\frac{\lambda}{am\|x_k - v_i\|^2}\right)^{\frac{1}{m-1}} = 1$$

$$\Rightarrow \left(\frac{\lambda}{am}\right)^{\frac{1}{m-1}} \sum_{i=1}^{c} \left(\frac{1}{\|x_k - v_i\|^2}\right)^{\frac{1}{m-1}} = 1 \tag{7.1}$$

$$\Rightarrow \left(\frac{\lambda}{am}\right)^{\frac{1}{m-1}} = \left(\sum_{i=1}^{c} \left(\frac{1}{\|x_k - v_i\|^2}\right)^{\frac{1}{m-1}}\right)^{-1}$$

After putting this value again in (7.1), we get

$$\mu_{ik} = \left[\sum_{j=1}^{C} \left(\frac{D_{ik}}{D_{jk}}\right)^{2/m-1}\right]^{-1}$$

where

$$D_{ik}{}^2 = ||x_k - v_i||^2$$

Derivation for possibilistic membership t_{ik}

For finding possibilistic membership, we differentiate J_{PFCM} with respect to t_{ik}:

$$\frac{\partial J_{\text{PFCM}}}{\partial t_{ik}} = b\eta t_{ik}^{\eta-1}||x_k - v_i||^2 - Y_i\eta(1 - t_{ik})^{\eta-1}$$

Set derivative to zero:

$$=> b\eta t_{ik}^{\eta-1}||x_k - v_i||^2 = \eta(1 - t_{ik})^{\eta-1}$$

Clear η and rearrange for t_{ik}:

$$=> \left(\frac{t_{ik}}{1 - t_{ik}}\right)^{\frac{1}{\eta-1}} = \frac{Y_i}{b||x_k - v_i||^2}$$

$$=> t_{ik} = \frac{1}{1 + \left(\frac{b||x_k - v_i||^2}{Y_i}\right)^{\frac{1}{\eta-1}}}$$

Derivation for cluster centre v_i

Differentiate J_{PFCM} with respect to v_i:

$$\frac{\partial J_{\text{PFCM}}}{\partial v_i} = \sum_{k=1}^{N} -2\left(a\mu_{ik}^m + bt_{ik}^{\eta}\right)(x_k - v_i)$$

Set derivative to zero:

$$=> \sum_{k=1}^{N} -2\left(a\mu_{ik}^m + bt_{ik}^{\eta}\right)(x_k - v_i) = 0$$

$$=> \sum_{k=1}^{N}\left(a\mu_{ik}^m + bt_{ik}^{\eta}\right)x_k = v_i\sum_{k=1}^{N}\left(a\mu_{ik}^m + bt_{ik}^{\eta}\right)$$

$$=> v_i = \frac{\sum_{k=1}^{N}\left(a\mu_{ik}^m + bt_{ik}^{\eta}\right)x_k}{\sum_{k=1}^{N}\left(a\mu_{ik}^m + bt_{ik}^{\eta}\right)}$$

Derive scale parameter Y_i

So, to find the optimal scale parameter Y_i, that minimizes the total penalized distance $||x_k - v_i||^2$ we equate the two contributions. Minimization suggests that the optimal Y_i balances the two terms;

$$\sum_{k=1}^{N} t_{ik}^{\eta} ||x_k - v_i||^2 \approx Y_i \sum_{k=1}^{N} (1 - t_{ik})^{\eta-1}$$

Solve for Y_i, we get

$$Y_i = \frac{\sum_{j=1}^{N} t_{ij}^{\eta} ||x_j - v_i||^2}{\sum_{j=1}^{N} (1 - t_{ij})^{\eta}}$$

PFCM:

```
a=1; %user defined constants that represent relative importance of fuzzy and possibilistic terms of the algorithm
b=1; %user defined constants that represent relative importance of fuzzy and possibilistic terms of the algorithm
m=2;
n=2; %possibilistic exponent
Choose number of cluster 'C' centre convenient for your data;
Take initial cluster centres as the centres obtained from FCM algorithm;

Start "For" loop for number of iterations (say r);
Take data set;
Find distance 'd' between cluster centres and data points;

Again start "For" loop for finding membership values iteratively as many times as number of data points.
Find membership values using;
Find value of 'μ_ik' using;
```

$$\mu_{ik} = \left[\sum_{j=1}^{C} \left(\frac{D_{ik}}{D_{jk}} \right)^{2/m-1} \right]^{-1}$$

```
End "For" loop for membership values;

In next step find value of 'Y_i' using;
```

$$Y_i = K \frac{\sum_{j=1}^{N} t_{ij}^{\eta} \left\| x_j - v_i \right\|^2}{\sum_{j=1}^{N} \left(1 - t_{ij}\right)^{\eta}}$$

Again start “For” loop for finding typicality values iteratively as many times as number of data points.

Find typicality values using;

$$t_{ik} = \frac{1}{1 + \left(\frac{b D_{ikA}^2}{Y_i}\right)^{\frac{1}{\eta - 1}}}$$

End the loop for typicality values;

In next step find value of cluster centres using;

$$v_i = \frac{\sum_{k=1}^{N} (a\mu_{ik}^m + bt_{ik}^{\eta}) x_k}{\sum_{k=1}^{N} (a\mu_{ik}^m + bt_{ik}^{\eta})}$$

In next step put the terminating condition for checking optimization of solution.

End the “For” loop for number of iterations ‘r’;

In next step find the value of objective function;

$$J_{\text{PFCM}} = \sum_{k=1}^{N} \sum_{i=1}^{C} (a\mu_{ik}^m + bt_{ik}^{\eta}) \times \|x_k - v_i\|^2 + \sum_{i=1}^{C} Y_i \sum_{k=1}^{N} (1 - t_{ik})^{\eta}$$

And plot the data points and centres on the graph for better analysis of data.

7.3 Python code

```
import numpy as np
import matplotlib.pyplot as plt

def pfcm(X, C=3, m=2, eta=2, a=1, b=1, max_iter=100, tol=1e-5, random_-
state=42):
    """
    Possibilistic Fuzzy C - Means (PFCM) clustering implementation.
    Parameters:
        X : np.ndarray of shape (N, d) - dataset
```

```
        C : int - number of clusters
        m : fuzziness exponent
        eta : possibilistic exponent
        a, b : user-defined weights for fuzzy and possibilistic terms
        max_iter : maximum number of iterations
        tol : tolerance for convergence
    Returns:
        v : final cluster centers
        mu : membership matrix
        t : typicality matrix
        obj_values : objective function history
    """
    np.random.seed(random_state)
    N, d = X.shape

    # Step 1: Initialize memberships (from random, or could use FCM result)
    mu = np.random.rand(C, N)
    mu /= np.sum(mu, axis=0, keepdims=True)

    # Initialize typicality matrix randomly
    t = np.random.rand(C, N)
    t /= np.sum(t, axis=0, keepdims=True)

    obj_values = []

    for iteration in range(max_iter):
        # Step 2: Compute distances
        v = np.dot((a * mu**m + b * t**eta), X) / (np.sum(a * mu**m + b * t**eta,
axis=1, keepdims=True) + 1e-8)
        D = np.zeros((C, N))
        for i in range(C):
            D[i] = np.linalg.norm(X - v[i], axis=1) + 1e-8 # add epsilon to
avoid division by zero

        # Step 3: Update membership values
        mu_new = np.zeros_like(mu)
        for k in range(N):
            for i in range(C):
                denom = np.sum((D[i, k] / D[:, k]) ** (2/(m-1)))
                mu_new[i, k] = 1 / (denom + 1e-8)
```

```
        # Step 4: Calculate Y_i for each cluster
        Y = np.zeros(C)
        K = 1 # usually K=1
        for i in range(C):
            num = np.sum(t[i]**eta * (D[i]**2))
            denom = np.sum((1 - t[i])**eta)
            Y[i] = K * num / (denom + 1e-8)

        # Step 5: Update typicality values
        t_new = np.zeros_like(t)
        for k in range(N):
            for i in range(C):
                power = (b * (D[i, k]**2) / (Y[i] + 1e-8)) ** (1/(eta - 1))
                t_new[i, k] = 1 / (1 + power)

        # Step 6: Compute objective function
        obj1 = np.sum((a * mu_new**m + b * t_new**eta) * (D**2))
        obj2 = np.sum([Y[i] * np.sum((1 - t_new[i])**eta) for i in range(C)])
        J = obj1 + obj2
        obj_values.append(J)

        # Step 7: Check convergence
        if np.linalg.norm(mu_new - mu) < tol and np.linalg.norm(t_new - t)
< tol:
            print(f'Converged at iteration {iteration+1}')
            break

        mu = mu_new
        t = t_new

    # Plot objective function over iterations
    plt.figure(figsize=(6, 4))
    plt.plot(obj_values, '-o')
    plt.title('Objective Function over Iterations (PFCM)')
    plt.xlabel('Iteration')
    plt.ylabel('J_PFCM')
    plt.grid(True)
    plt.show()

    # Plot final clustering result
    plt.figure(figsize=(6, 6))
    cluster_labels = np.argmax(mu, axis=0)
```

```
    plt.scatter(X[:, 0], X[:, 1], c=cluster_labels, cmap='viridis',
alpha=0.6)
    plt.scatter(v[:, 0], v[:, 1], c='red', s=200, marker='X', label='-
Centers')
    plt.title('PFCM Clustering Result')
    plt.legend()
    plt.show()

    return v, mu, t, obj_values

# Example usage
if __name__ == "__main__":
    # Generate toy data
    np.random.seed(0)
    X = np.vstack([
        np.random.randn(50, 2) + [2, 2],
        np.random.randn(50, 2) + [-2, -2],
        np.random.randn(50, 2) + [4, -3]
    ])
    centers, memberships, typicality, history = pfcm(X, C=3, max_iter=100,
a=1, b=1)
    print("Final cluster centers:\n", centers)
```

7.4 R-code

```
# Load required library
library(ggplot2)
pfcm <- function(X, C=3, m=2, eta=2, a=1, b=1, max_iter=100, tol=1e-5,
seed=42) {
    set.seed(seed)
    N <- nrow(X)
    d <- ncol(X)

    # Step 1: Initialize membership and typicality randomly
    mu <- matrix(runif(C * N), nrow=C)
    mu <- mu / matrix(colSums(mu), nrow=C, ncol=N, byrow=TRUE)

    t <- matrix(runif(C * N), nrow=C)
    t <- t / matrix(colSums(t), nrow=C, ncol=N, byrow=TRUE)

    obj_values <- c()
```

```
for (iter in 1:max_iter) {
  # Step 2: Compute cluster centers
  numerator <- a * (mu^m) %*% as.matrix(X) + b * (t^eta) %*% as.matrix(X)
  denominator <- rowSums(a * (mu^m) + b * (t^eta))
  v <- numerator / denominator

  # Step 3: Compute distances
  D <- matrix(0, nrow=C, ncol=N)
  for (i in 1:C) {
    diff <- sweep(X, 2, v[i,])
    D[i, ] <- sqrt(rowSums(diff^2)) + 1e-8
  }

  # Step 4: Update membership values
  mu_new <- matrix(0, nrow=C, ncol=N)
  for (k in 1:N) {
    for (i in 1:C) {
    denom <- sum((D[i, k] / D[, k])^(2/(m-1)))
    mu_new[i, k] <- 1 / (denom + 1e-8)
    }
  }

  # Step 5: Calculate Y_i
  Y <- numeric(C)
  K <- 1
  for (i in 1:C) {
    num <- sum(t[i,]^eta * (D[i,]^2))
    denom <- sum((1 - t[i,])^eta)
    Y[i] <- K * num / (denom + 1e-8)
  }

  # Step 6: Update typicality values
  t_new <- matrix(0, nrow=C, ncol=N)
  for (i in 1:C) {
    for (k in 1:N) {
      power <- (b * (D[i,k]^2) / (Y[i]+1e-8))^(1/(eta-1))
      t_new[i,k] <- 1 / (1 + power)
    }
  }

  # Step 7: Compute objective function
  obj1 <- sum((a * mu_new^m + b * t_new^eta) * (D^2))
```

```
    obj2 <- sum(Y * apply((1 - t_new)^eta, 1, sum))
    J <- obj1 + obj2
    obj_values <- c(obj_values, J)

    # Step 8: Check convergence
    if (max(abs(mu_new - mu)) < tol && max(abs(t_new - t)) < tol) {
     cat(sprintf("Converged at iteration %d\n", iter))
     break
    }

    mu <- mu_new
    t <- t_new
  }

  # Plot objective function
  plot(obj_values, type="o", col="blue", xlab="Iteration", ylab="Objec-
tive Function (J_PFCM)",
       main="PFCM: Objective Function over Iterations")

  # Plot clustering result
  cluster_labels <- apply(mu, 2, which.max)
  df <- data.frame(X1=X[,1], X2=X[,2], cluster=as.factor(cluster_labels))
  centers_df <- data.frame(v)
  colnames(centers_df) <- c("X1", "X2")

  ggplot(df, aes(x=X1, y=X2, color=cluster)) +
    geom_point(size=2, alpha=0.7) +
    geom_point(data=centers_df, aes(x=X1, y=X2), color='red', size=5,
shape=4, stroke=2) +
    ggtitle("PFCM Clustering Result") +
    theme_minimal()

  return(list(centers=v, membership=mu, typicality=t, objective=obj_-
values))
}

# Example usage
set.seed(0)
X <- rbind(
  matrix(rnorm(100, mean=2), ncol=2),
  matrix(rnorm(100, mean=-2), ncol=2),
  matrix(rnorm(100, mean=4), ncol=2)
```

```
)
result <- pfcm(X, C=3, max_iter=100)
print("Final cluster centers:")
print(result$centers)
```

7.5 Solved example

Example 1

The study investigates 15 CSK IPL players' stats including batting averages, over rates, and selection positions shown in Table 7.1. Employing the PFCM clustering technique, we want to categorize the players into three unique clusters. Our analysis groups players into batters, all-rounders, and fast bowlers. Through this technique, we discover the strongest bowling all-rounder performers within our dataset.

Table 7.1: CSK team members' batting and bowling statistics along with their early designations.

Player	Batting average	Bowl average	Initial class
MM Ali	24.96	22.78	Batsman
S Anirudha	19	50	Bowler
R Ashwin	9.5	24.22	Bowler
S Badrinath	30.65	50	Batsman
L Balaji	3.66	26.09	Bowler
DE Bollinger	21	18.72	Batsman
DJ Bravo	25.1	22.47	Batsman
DL Chahar	10.83	28.2	Bowler
SM Curran	22	30.54	Batsman
MS Dhoni	40.53	50	Batsman
F du Plessis	35.33	50	Batsman
RD Gaikwad	39.36	50	Batsman
MS Gony	12.5	30.73	Bowler
Harbhajan Singh	7.5	25.3	Bowler

Research inquiries:

(i) **Classification of players:**
What is the use of PFCM clustering for categorizing players into three distinct categories based on their batting and bowling averages: Our system identifies players into three separate groups, which are batsmen, bowling all-rounders, and bowlers.

(ii) **Determination of optimal bowling all-rounders:**
What are the profiles of players in our sample who perform well both as bowlers and batsmen and how do they stack up versus other player types?

(iii) **Cluster analysis:**
How does each performance segment appear different through its bowling and batting statistics?
The discovered clusters show how much they match the player categories from step one.

(iv) **Performance analysis:**
By using PFCM clustering, we can find player performance patterns that we cannot see through simple data review.

Solution:

Step 1: Consider batting average = X1 and bowling average = X2.

X1	24.96	19	9.5	30.65	3.66	21	25.1	10.83	22	40.53	35.33	39.36	12.5	7.5
X2	22.78	50	24.22	50	26.09	18.72	22.47	28.2	30.54	50	50	50	30.73	25.3

Step 2: The algorithm is initialized by assigning initial cluster centres as shown below.

v1	17.5540	34.6180
v2	19.7325	24.9825
v3	27.0480	41.2060

Step 3: Initial membership values for initializing the clustering algorithm.

U1	0.009	0.233	0.954	0.019	0.914	0.113	0.012	0.972	0.213	0.019	0.000	0.012	0.851	0.976
U2	0.003	0.566	0.006	0.957	0.017	0.020	0.004	0.004	0.070	0.953	0.996	0.969	0.025	0.004
U3	0.988	0.200	0.040	0.024	0.068	0.867	0.984	0.024	0.717	0.029	0.000	0.019	0.125	0.020

Step 3: Take $m = 2$.

U1^2	U2^2	U3^2	X1*U1^2	X2*U2^2	X1*U3^2	X2*U2^2	X1*U3^2	X2*U3^2
8.91E-05	9.32E-06	0.975172	0.002224	0.002029637	0.000233	0.000212	24.34029	22.21441982
0.05444	0.320807	0.040111	1.034363	2.722008372	6.095334	16.04035	0.762111	2.005554055
0.909592	3.58E-05	0.001623	8.641121	22.03031173	0.00034	0.000868	0.01542	0.039312236
0.00035	0.91563	0.000596	0.010713	0.01747694	28.06405	45.78148	0.018276	0.029813494
0.83604	0.000306	0.004645	3.059906	21.8122833	0.00112	0.007985	0.017	0.121184326
0.012696	0.000417	0.75152	0.266622	0.237674597	0.008756	0.007806	15.78191	14.06844884
0.000152	1.62E-05	0.96755	0.003817	0.003417342	0.000407	0.000364	24.28549	21.74083743
0.944111	1.74E-05	0.000585	10.22472	26.62392766	0.000188	0.00049	0.006331	0.01648642
0.045331	0.004912	0.514096	0.997284	1.384412102	0.108059	0.150005	11.31012	15.70050067
0.000353	0.907604	0.000814	0.014316	0.017660613	36.78521	45.38022	0.032974	0.040678988
0	0.991394	0	0	0	35.02595	49.56971	0	0
0.000154	0.938844	0.000348	0.006059	0.00769727	36.95291	46.94221	0.013694	0.017396339
0.723552	0.000611	0.01554	9.044404	22.23476191	0.007638	0.018778	0.194256	0.47755799
0.951905	1.5E-05	0.000419	7.139289	24.08320183	0.000112	0.000379	0.003144	0.010607031
4.478766	**4.080619**	**3.273019**	**40.44484**	**121.1768633**	**143.0503**	**203.9009**	**76.78103**	**76.48279764**

Step 4: Final cluster centres.

v1	9.1884	26.8611
v2	5.4536	49.9326
v3	23.6877	23.0632

Example 2

An organization has warehouses for life saving drugs at six different locations, whose coordinates are given in the following table. The organization is in the process of locating one more critical central warehouse which will distribute drugs to all the existing warehouses on emergency request. Find the number of optimum locations of the new facility (warehouse) based on the PFCM clustering concept.

Table 7.2: Thirty warehouses whose coordinates are summarized.

Existing warehouse number	Coordinates of centroids
1	(2.5,3.5)
2	(2.7,3.6)
3	(2.2,3.2)
4	(3.0,3.0)
5	(2.7,3.1)
6	(2.3,3.5)
7	(3.1,3.0)
8	(3.3,2.9)
9	(3.0,3.5)
10	(2.9, 2.9)
11	(6.0,7.2)
12	(6.5,7.0)
13	(5.9,6.9)
14	(6.3,7.3)
15	(5.9,7.0)
16	(6.8,7.0)
17	(6.5,6.9)
18	(6.2,7.1)
19	(7.2,7.2)
20	(6.6,7.1)
21	(11.1,13.5)
22	(12.2,12.5)
23	(11.5,11.9)

Solution:

The data are shown in Table 7.2. Assume that we want to determine a fuzzy pseudo-partition with three clusters (i.e. $c = 3$). Assume further that we choose $m = 2$ and $||.||$ is the Euclidean distance. The algorithm is initialized by randomly assigning membership values and typicality values across three cluster centres, forming the initial fuzzy and possibilistic partitions, respectively, as $T^{(0)} = [T_1, T_2, T_3]$. Table 7.3–7.7 shows the results obtained at different iterations.

Table 7.3: Cluster centres and the corresponding membership function values of matrix *T*.

Input data		Membership function for fuzzy clustering		
X_1	X_2	T_1	T_2	T_3
2.5	3.5	.9	.05	.05
2.7	3.6	.89	.055	.055
2.2	3.2	.88	.06	.06
3.0	3.0	.87	.065	.065
2.7	3.1	.86	.07	.07
2.3	3.5	.84	.09	.07
3.1	3.0	.81	.11	.08
3.3	2.9	.8	.1	.1
3.0	3.5	.79	.11	.1
2.9	2.9	.75	.15	.1
6.0	7.2	.05	.9	.05
6.5	7.0	.055	.89	.055
5.9	6.9	.06	.88	.06
6.3	7.3	.065	.87	.065
5.9	7.0	.07	.86	.07
6.8	7.0	.09	.84	.07
6.5	6.9	.11	.81	.08
6.2	7.1	.1	.8	.1
7.2	7.2	.11	.79	.1
6.6	7.1	.15	.75	.1
11.1	13.5	.05	.05	.9
12.2	12.5	.055	.055	.89
11.5	11.9	.06	.06	.88
13.2	13.3	.065	.065	.87
11.5	12.5	.07	.07	.86
12.0	15.0	.07	.09	.84
13.2	13.2	.08	.11	.81
11.2	11.9	.1	.1	.8
12.9	13.7	.1	.11	.79
13.9	13.0	.1	.15	.75

Iteration 1:

Table 7.4: Cluster centres obtained after the first iteration.

v1	2.7862	3.2344
v2	6.3767	7.0673
v3	12.2425	12.9787

Iteration 2:

Table 7.5: Cluster centres obtained after the second iteration.

v1	2.7786	3.2175
v2	6.3855	7.0712
v3	12.2836	13.0080

Due to space and readability considerations, it is impractical to present the complete set of iterative results. Therefore, the following tables provide a summary of the cluster centres obtained during the penultimate (i.e. 15th) and final (i.e. 16th) iterations, which effectively illustrate the convergence behaviour of the algorithm.

Iteration 11:

Table 7.6: Cluster centres obtained after the 11th iteration.

v1	2.7788	3.2140
v2	6.3885	7.0712
v3	12.3062	13.0231

Iteration 12:

Table 7.7: Cluster centres and the corresponding last iteration.

v1	2.7789	3.2140
v2	6.3884	7.0713
v3	12.3061	13.0230

Then, the algorithm stops after 13 iterations and we obtain the pseudo-partition defined in Table 2.5. The three cluster centres are $V_1 = (2.7789, 3.2140)$, $V_2 = (6.3884, 7.0713)$, and $V_3 = (12.3061, 13.0231)$, and $J_{\text{PFCM}} = 24.0782$.

7.6 Problem set

Problem 1

This study observes five students based on their **average sleep hours** and **academic performance (out of 100)**. One student shows poor academics despite high sleep, possibly due to a lack of focus. **PFCM** is used to group students into:

- **High performers**
- **Low performers**
- **Moderate performers**

PFCM helps account for uncertainty in the data by assigning both **fuzzy memberships** and **possibilistic typicalities**.

Table 1: Sleep versus academic performance.

Student	Sleep hours	Performance score
A	6.5	85
B	8.0	88
C	4.5	40
D	7.0	80
E	9.5	30 (ambiguous)

Research inquiries:

(i) **Combined role of membership and typicality:** How do fuzzy membership and typicality together help understand ambiguous students like E?

(ii) **Cluster interpretation:** What do the cluster centres indicate about the balance between rest and performance?

(iii) **Pattern discovery:** How does PFCM provide insights that FCM or PCM alone may miss?

(iv) **Handling overlap and outliers:** Which students fall into fuzzy zones or are low in typicality?

Problem 2

We examine five students based on **class participation, homework completion**, and **attendance rate**. Some students perform well only in certain aspects. PFCM clusters them into:
- **Highly engaged students**
- **Passive but present**
- **Active but irregular**

By combining fuzzy and possibilistic analysis, PFCM reveals **participation styles** more clearly than any one measure.

Table 2: Participation metrics.

Student	Participation (%)	Homework (%)	Attendance (%)
A	90	95	85
B	40	30	90
C	85	80	55
D	35	50	60
E	88	20	70 (conflicting)

Research inquiries:
(i) **Partial belonging detection:** Which students belong partially to multiple clusters, and how is this seen through membership?
(ii) **Typicality reliability:** Which students have high membership but low typicality, and what does this reveal?
(iii) **Cluster centre analysis:** What do the cluster centres suggest about study behaviours?
(iv) **Application in school reports:** How can schools use PFCM results to guide counselling or support?

Problem 3

This study classifies customers based on their behaviour: **Items viewed, items purchased, returns made**, and **customer ratings**. PFCM is used to create clusters:
- **Genuine buyers**
- **Window shoppers**
- **Frequent returners**

The dual nature of PFCM helps identify **uncertain customer types** more effectively than hard clustering.

Table 3: E-commerce user data.

User	Views	Purchases	Returns	Rating (/5)
U1	50	10	0	4.5
U2	45	2	6	2.5
U3	55	8	1	4.0
U4	80	0	0	3.0
U5	20	9	8	4.8 (inconsistent)

Research inquiries:

(i) **Dual analysis strength:** How does PFCM help in handling users like U5 who show both high returns and good ratings?

(ii) **Fuzzy-zone customers:** Which customers fall into multiple behaviour groups based on their memberships?

(iii) **Typicality interpretation:** How does typicality help determine reliable representatives of each customer type?

(iv) **E-commerce decision making:** How can businesses act on PFCM-based clusters to improve retention and reduce returns?

References

[1] U. Fayyad, G. Piatetsky-Shapiro, P. S.-A. I. magazine, and undefined 1996, "From data mining to knowledge discovery in databases," *AI magazine*, vol. 17, no. 3, pp. 37–37.

[2] G. J. Oyewole, and G. A. Thopil, "Data clustering: application and trends," *Artificial Intelligence Review*, vol. 56, no. 7, pp. 6439–6475, Jul. 2023, doi: 10.1007/s10462-022-10325-y.

[3] D. T. L.-T. et adaptation de Thierry Vallaud and undefined 2005, "An introduction to data mining," *ebookyab.ir*, 2014, [Online]. Available: https://www.ebookyab.ir/files/sample – Solution Manual for Discovering Knowledge In Data 2nd edition Daniel Larose.pdf

[4] M. K. Gupta, and P. Chandra, "A comprehensive survey of data mining," *International Journal of Information Technology (Singapore)*, vol. 12, no. 4, pp. 1243–1257, Dec. 2020, doi: 10.1007/S41870-020-00427-7/METRICS.

[5] G. De Tre, A. Hallez, and A. Bronselaer, "Performance optimization of object comparison," *International Journal of Intelligent Systems*, vol. 29, no. 2, pp. 495–524, 2014, doi: 10.1002/int.

[6] R. Kumar, "A Comparative Study of Fuzzy Optimization through Fuzzy Number," vol. 1, no. 1, pp. 1–14, 2021.

[7] A. L. N. Fred, A. K. J.-I. Transactions on pattern analysis, and undefined 2005, "Combining multiple clusterings using evidence accumulation," in *IEEE Transactions on Pattern Analysis and Machine Intelligence*, vol. 27, no. 6, pp. 835–850, June 2005, doi: 10.1109/TPAMI.2005.113.

[8] S. Miyamoto, H. Ichihashi, K. Honda, and H. Ichihashi, *Algorithms for fuzzy clustering*. 2008. [Online]. Available: https://link.springer.com/content/pdf/10.1007/978-3-540-78737-2.pdf

[9] W. Pedrycz, and G. Vulkovich, "Fuzzy clustering with supervision," *Pattern Recognition*, vol. 37, no. 7, pp. 1339–1349, 2004, doi: 10.1016/j.patcog.2003.11.005.

[10] A. Pérez-Suárez, J. F. Martínez-Trinidad, and J. A. Carrasco-Ochoa, "A review of conceptual clustering algorithms," Aug. 15, 2019, *Springer Netherlands*. doi: 10.1007/s10462-018-9627-1.

[11] A. Lotfi, P. Moradi, and H. Beigy, "Density peaks clustering based on density backbone and fuzzy neighborhood," *Pattern Recognition*, vol. 107, Nov. 2020, doi: 10.1016/j.patcog.2020.107449.

[12] I. Kumar, H. S. Bhadauria, J. Virmani, S. T.-B. Engineering, and undefined 2017, "A classification framework for prediction of breast density using an ensemble of neural network classifiers," *Biocybernetics and Biomedical Engineering*, vol. 37, no.1, pp. 217–228.

[13] C. X. Ren, J. Z. Li, P. Ge, and X. L. Xu, "Deep metric learning via subtype fuzzy clustering," *Pattern Recognition*, vol. 90, pp. 210–219, Jun. 2019, doi: 10.1016/j.patcog.2019.01.037.

[14] R. Kumar, V. Joshi, G. Dhiman, and W. Viriyasitavat, "An improved exponential metric space approach for C-mean clustering analysing," *Expert Systems*, p. e12896, 2021, doi: 10.1111/EXSY.12896.

[15] S. Broumi, and F. Smarandache, *New distance and similarity measures of interval neutrosophic sets*. 2014. [Online]. Available: https://books.google.com/books?hl=en&lr=&id=PsU7DwAAQBAJ&oi=fnd&pg=PA17&dq=New+Distance+and+Similarity+Measures+of+Interval+Neutrosophic+Sets&ots=yoXb-5pNk3&sig=PVbwCcMBnlIg6ldZ-4wzvYZK2_s

[16] H. Motallebi, R. Nasihatkon, and M. Jamshidi, "A local mean-based distance measure for spectral clustering," *Pattern Analysis and Applications*, vol. 25, no. 2, pp. 351–359, May 2022, doi: 10.1007/s10044-021-01040-5.

[17] Z. Niu, Z. Yin, X. K.-2012 F. I. C. on, and undefined 2012, "Sentiment classification for microblog by machine learning," *ieeexplore.ieee.org*, [Online]. Available: https://ieeexplore.ieee.org/abstract/document/6300492/?casa_token=T9z0iFTjT9wAAAAA:Qc6Dq17-4X7qt_lqn1ibBnVm_xCWGvlIl52wsEzCHbp3WaWlbGerjceYPIOdkvJDeByc_1PZEAry

[18] S. A. L.-I. J. of Engineering and undefined 2017, "Oil Reservoirs Classification Using Fuzzy Clustering (RESEARCH NOTE)," *ije.ir*, [Online]. Available: https://www.ije.ir/article_73019.html

https://doi.org/10.1515/9783111494272-008

[19] H. Yu, S. Xie, J. Fan, R. Lan, and B. Lei, "Mahalanobis-Kernel Distance-based Suppressed Possibilistic C-Means Clustering Algorithm for Imbalanced Image Segmentation," *IEEE Transactions on Fuzzy Systems*, pp. 1–14, 2024, doi: 10.1109/TFUZZ.2024.3405497.

[20] P.-N. Tan, M. Steinbach, and V. Kumar, *Introduction to Data Mining (New International Edition)*. 2013.

[21] M. A. Fligner, and J. S. Verducci, "Distance Based Ranking Models," *Journal of the Royal Statistical Society: Series B (Methodological)*, 1986, doi: 10.1111/j.2517-6161.1986.tb01420.x.

[22] J. S. Lee, and C. L. Teng, "An Enhanced Hierarchical Clustering Approach for Mobile Sensor Networks Using Fuzzy Inference Systems," *IEEE Internet Things Journal*, vol. 4, no. 4, pp. 1095–1103, 2017, doi: 10.1109/JIOT.2017.2711248.

[23] N. Gillet, R. J. Vallerand, S. Amoura, and B. Baldes, "Influence of coaches' autonomy support on athletes' motivation and sport performance: A test of the hierarchical model of intrinsic and extrinsic motivation," *Psychology of Sport and Exercise*, 2010, doi: 10.1016/j.psychsport.2009.10.004.

[24] D. T. Anderson, J. C. Bezdek, M. Popescu, and J. M. Keller, "Comparing fuzzy, probabilistic, and possibilistic partitions," *IEEE Transactions on Fuzzy Systems*, vol. 18, no. 5, pp. 906–918, 2010, doi: 10.1109/TFUZZ.2010.2052258.

[25] F. Informa, W. R. Number, M. House, M. Street, and J. C. Dunn, "Well-Separated Clusters and Optimal Fuzzy Partitions Well-Separated Clusters and Optimal Fuzzy Partitions," no. June 2013, pp. 37–41, 2008.

[26] J. C. Dunn, "Well-Separated Clusters and Optimal Fuzzy Partitions," https://doi.org/10.1080/01969727408546059, vol. 4, no. 1, pp. 95–104, 2008, doi: 10.1080/01969727408546059.

[27] J. C. Bezdek, R. Ehrlich, and W. Full, "FCM: The fuzzy c-means clustering algorithm," *Computational Geosciences*, vol. 10, no. 2–3, pp. 191–203, 1984, doi: 10.1016/0098-3004(84)90020-7.

[28] P. Kumar, and D. Sirohi, "Comparative Analysis of FCM and HCM Algorithm on Iris Data Set," *International Journal of Computer Applications (0975–8887)*, vol. 5, no. 2, pp. 975–8887, 2010.

[29] L. Barbosa, J. F.-C. 2010: Posters, and undefined 2010, "Robust sentiment detection on twitter from biased and noisy data," *aclanthology.org*, pp. 36–44, 2010, [Online]. Available: https://aclanthology.org/C10-2005.pdf

[30] J. Wang, Y. Wang, F. Nie, and X. Li, "Discriminative projection fuzzy K-Means with adaptive neighbors," *Pattern Recognition Letters*, vol. 176, pp. 21–27, Dec. 2023, doi: 10.1016/J.PATREC.2023.10.008.

[31] D. Lei, Q. Zhu, J. Chen, H. Lin, and P. Yang, "Automatic k-means clustering algorithm for outlier detection," *Lecture Notes in Electrical Engineering*, vol. 154 LNEE, pp. 363–372, 2012, doi: 10.1007/978-1-4471-2386-6_47/COVER.

[32] T. M. Ghazal *et al.*, "Performances of K-Means Clustering Algorithm with Different Distance Metrics", doi: 10.32604/iasc.2021.019067.

[33] P. Sangwan, R. Kumar, A. Kaur, and G. Dhiman, "A Comprehensive Study of Various Fuzzy C-Means Clustering Algorithms," in *AI-Enabled Multiple-Criteria Decision-Making Approaches for Healthcare Management*, IGI Global, 2022, pp. 161–180.

[34] S. R. Kannan, R. Devi, S. Ramathilagam, and K. Takezawa, "Effective FCM noise clustering algorithms in medical images," *Computers in Biology and Medicine [Internet]*, vol. 43, no. 2, pp. 73–83, Feb. 2013, doi: 10.1016/j.compbiomed.2012.10.002.

[35] R. Ceylan, and Y. Özbay, "Comparison of FCM, PCA and WT techniques for classification ECG arrhythmias using artificial neural network," *Expert Systems with Applications*, vol. 33, no. 2, pp. 286–295, 2007, doi: 10.1016/J.ESWA.2006.05.014.

[36] S. R. Kannan, R. Devi, S. Ramathilagam, and K. Takezawa, "Effective FCM noise clustering algorithms in medical images," *Computers in Biology and Medicine [Internet]*, vol. 43, no. 2, pp. 73–83, Feb. 2013, doi: 10.1016/j.compbiomed.2012.10.002.

[37] C. Qiao, C. Wu, ·Changxing Li, and J. Wang, "Robust fuzzy clustering algorithms integrating membership guided image filtering," *Signal, Image Video Process*, vol. 16, no. 7, pp. 1851–1859, Oct. 2022, doi: 10.1007/s11760-022-02144-z.

[38] S. Zhou, D. Li, Z. Zhang, and R. Ping, "A New Membership Scaling Fuzzy C-Means Clustering Algorithm," *IEEE Transactions on Fuzzy Systems*, vol. 29, no. 9, pp. 2810–2818, Sep. 2021, doi: 10.1109/TFUZZ.2020.3003441.

[39] R. Krishnapuram and J. M. Keller, "A Possibilistic Approach to Clustering," *IEEE Transactions on Fuzzy Systems*, vol. 1, no. 2, pp. 98–110, 1993, doi: 10.1109/91.227387.

[40] H. J. Fu, X. H. Wu, H. P. Mao, and B. Wu, "Fuzzy entropy clustering using possibilistic approach," in *Procedia Engineering*, 2011, pp. 1993–1997. doi: 10.1016/j.proeng.2011.08.372.

[41] H. Tanaka, I. Hayashi, J. W.-E. J. of Operational, and undefined 1989, "Possibilistic linear regression analysis for fuzzy data," *Elsevier*, [Online]. Available: https://www.sciencedirect.com/science/article/pii/0377221789904311

[42] R. Krishnapuram, J. M. K.-I. transactions on Fuzzy, and undefined 1996, "The possibilistic c-means algorithm: insights and recommendations," *ieeexplore.ieee.org*, [Online]. Available: https://ieeexplore.ieee.org/abstract/document/531779/

[43] M. S. Yang, and Y. Nataliani, "Robust-learning fuzzy c-means clustering algorithm with unknown number of clusters," *Pattern Recognition*, vol. 71, pp. 45–59, Nov. 2017, doi: 10.1016/j.patcog.2017.05.017.

[44] M. Cimino, and G. Frosini, "On the noise distance in robust fuzzy c-means," *International Journal on Engineering, Computing and Technology*, no. December, pp. 361–364, 2007, [Online]. Available: http://www2.ing.unipi.it/~r000099/publications/cimino_pub19.pdf

[45] S. Al-gawwam, M. B. Information, and undefined 2018, "Robust eye blink detection based on eye landmarks and Savitzky–Golay filtering," *Information*, vol. 9, pp. 93. https://doi.org/10.3390/info9040093

[46] Q. Wang, X. Wang, C. Fang, and W. Yang, "Robust fuzzy c-means clustering algorithm with adaptive spatial & intensity constraint and membership linking for noise image segmentation," *Applied Soft Computing*, vol. 92, p. 106318, Jul. 2020, doi: 10.1016/J.ASOC.2020.106318.

[47] Pooja, R. Kumar, W. Viriyasitavat, K. Yadav, and G. Dhiman, "Analysis of Clustering Algorithms for Facility Location Allocation Problems," pp. 597–605, 2023, doi: 10.1007/978-981-19-9228-5_51.

[48] P. Sangwan, R. Kumar, Y. Sharma, D. G. Bhosale, and C. D. Prasad, "Application of computing in recognition of input design factors for vapour-grown carbon nanofibers through fuzzy cluster analysis," *International Journal on Interactive Design and Manufacturing*, pp. 1–11, Oct. 2023, doi: 10.1007/S12008-023-01547-7/METRICS.

[49] A. Singh and A. Kumar, "Fuzzy-based approach to incorporate spatial constraints in possibilistic c-means algorithm for remotely sensed imagery," *International Journal of Intelligent Information and Database Systems*, vol. 13, no. 2–4, pp. 307–318, 2020, doi: 10.1504/IJIIDS.2020.109459.

[50] H. Timm, C. Borgelt, C. Döring, and R. Kruse, "An extension to possibilistic fuzzy cluster analysis," *Fuzzy Sets and Systems*, vol. 147, no. 1, pp. 3–16, 2004, doi: 10.1016/j.fss.2003.11.009.

[51] N. R. Pal, K. Pal, J. M. Keller, and J. C. Bezdek, "A possibilistic fuzzy c-means clustering algorithm," *IEEE Transactions on Fuzzy Systems*, vol. 13, no. 4, pp. 517–530, Aug. 2005, doi: 10.1109/TFUZZ.2004.840099.

[52] J. Chen, H. Zhang, D. Pi, M. Kantardzic, Q. Yin, and X. Liu, "A weight possibilistic fuzzy C-means clustering algorithm," *Science Program*, vol. 2021, 2021, doi: 10.1155/2021/9965813.

[53] N. R. Pal, K. Pal, J. M. K.- I. Transactions on fuzzy . . ., and undefined 2005, "A possibilistic fuzzy c-means clustering algorithm," in *IEEE Transactions on Fuzzy Systems*, vol. 13, no. 4, pp. 517–530, Aug. 2005, doi: 10.1109/TFUZZ.2004.840099.

[54] M.-B. Naghi, L. Kovácsóbuda, and L. Szilágyi, "A generalized fuzzy-possibilistic c-means clustering algorithm," *Acta Universitatis Sapientiae, Informatica*, vol. 15, pp. 404–431, 2023, doi: 10.2478/ausi-2023-0023.

[55] R. N. Dave, and S. Sen, "Noise clustering algorithm revisited," *Annual Conference of the North American Fuzzy Information Processing Society – NAFIPS*, no. 1, pp. 199–204, 1997, doi: 10.1109/nafips.1997.624037.

[56] R. N. Dave, "Characterization and detection of noise in clustering," *Pattern Recognition Letters*, vol. 12, no. 11, pp. 657–664, 1991, doi: 10.1016/0167-8655(91)90002-4.

Index

https://doi.org/10.1515/9783111494272-009

De Gruyter Series on the Applications of Mathematics in Engineering and Information Sciences

Already published in the series

Volume 22
Metaverse and Digital Twins: Blockchain and Healthcare Technology
Bhupinder Singh, Rishabha Malviya, Christian Kaunert (Eds.)
ISBN 978-3-11-163629-0, e-ISBN (PDF) 978-3-11-164689-3,
e-ISBN (EPUB) 978-3-11-164464-6

Volume 21
Reengineering Cyber Security Process: Quality Management
Mehrdad Sepehri Sharbaf
ISBN 978-3-11-144382-9, e-ISBN (PDF) 978-3-11-144463-5,
e-ISBN (EPUB) 978-3-11-144480-2

Volume 20
Digital Blockchain: Big Data, Artificial Intelligence, and Virtual Reality in Healthcare Ecosystem
Rishabha Malviya, Shristy Verma, Sonali Sundram, Harshil Shah
ISBN 978-3-11-157386-1, e-ISBN (PDF) 978-3-11-157428-8,
e-ISBN (EPUB) 978-3-11-157485-1

Volume 19
Artificial Intelligence for Healthcare: Machine Learning and Diagnostics
Rishabha Malviya, Shivam Rajput, Mukesh Roy, Irfan Ahmad, Saurabh Srivastava
ISBN 978-3-11-154758-9, e-ISBN (PDF) 978-3-11-154972-9,
e-ISBN (EPUB) 978-3-11-154984-2

Volume 18
Operational Perspective of Modeling System Reliability: Research Tools for System Dynamics
Deepti Aggrawal, Adarsh Anand, Yoshinobu Tamura, Mohini Agarwal (Eds.)
ISBN 978-3-11-147540-0, e-ISBN (PDF) 978-3-11-147610-0,
e-ISBN (EPUB) 978-3-11-147713-8

Volume 17
Algorithms: Big Data, Optimization Techniques, Cyber Security
Sushil C. Dimri, Abhay Saxena, Bhuvan Unhelkar, Akshay Kumar (Eds.)
ISBN 978-3-11-122800-6, e-ISBN (PDF) 978-3-11-122915-7,
e-ISBN (EPUB) 978-3-11-122963-8

Volume 16
Distributed Transfer Function Method: One-Dimensional Problems in Engineering
Bingen Yang, Kyoungrae Noh
ISBN 978-3-11-075854-2, e-ISBN (PDF) 978-3-11-075893-1,
e-ISBN (EPUB) 978-3-11-075900-6

Volume 15
Machine Learning for Cyber Security
Malik Preeti, Nautiyal Lata, Ram Mangey (Eds.)
ISBN 978-3-11-076673-8, e-ISBN (PDF) 978-3-11-076674-5,
e-ISBN (EPUB) 978-3-11-076676-9

Volume 14
Multiple Criteria Decision-Making Methods: Applications for Managerial Discretion
Mohini Agarwal, Adarsh Anand, Deepti Aggrawal
ISBN 978-3-11-074356-2, e-ISBN (PDF) 978-3-11-074363-0,
e-ISBN (EPUB) 978-3-11-074374-6

Volume 13
Integral Transforms and Applications
Nita H. Shah, Monika K. Naik
ISBN 978-3-11-079282-9, e-ISBN (PDF) 978-3-11-079285-0,
e-ISBN (EPUB) 978-3-11-079292-8

Volume 12
Noise Filtering for Big Data Analytics
Souvik Bhattacharyya, Koushik Ghosh (Eds.)
ISBN 978-3-11-069709-4, e-ISBN (PDF) 978-3-11-069721-6,
e-ISBN (EPUB) 978-3-11-069726-1

Volume 11
Artificial Intelligence for Signal Processing and Wireless Communication
Abhinav Sharma, Arpit Jain, Ashwini Kumar Arya, Mangey Ram (Eds.)
ISBN 978-3-11-073882-7, e-ISBN (PDF) 978-3-11-073465-2,
e-ISBN (EPUB) 978-3-11-073472-0

Volume 10
Meta-heuristic Optimization Techniques: Applications in Engineering
Anuj Kumar, Sangeeta Pant, Mangey Ram, Om Yadav (Eds.)
ISBN 978-3-11-071617-7, e-ISBN (PDF) 978-3-11-071621-4,
e-ISBN (EPUB) 978-3-11-071625-2

Volume 9
Linear Integer Programming: Theory, Applications, Recent Developments
Elias Munapo, Santosh Kumar
ISBN 978-3-11-070292-7, e-ISBN (PDF) 978-3-11-070302-3
e-ISBN (EPUB) 978-3-11-070311-5

Volume 8
Mathematics for Reliability Engineering: Modern Concepts and Applications
Mangey Ram, Liudong Xing (Eds.)
ISBN 978-3-11-072556-8, e-ISBN (PDF) 978-3-11-072563-6
e-ISBN (EPUB) 978-3-11-072559-9

Volume 7
Mathematical Fluid Mechanics: Advances on Convection Instabilities and Incompressible Fluid Flow
B. Mahanthesh (Ed.)
ISBN 978-3-11-069603-5, e-ISBN (PDF) 978-3-11-069608-0
e-ISBN (EPUB) 978-3-11-069612-7

Volume 6
Distributed Denial of Service Attacks: Concepts, Mathematical and Cryptographic Solutions
Rajeev Singh, Mangey Ram (Eds.)
ISBN 978-3-11-061675-0, e-ISBN (PDF) 978-3-11-061975-1
e-ISBN (EPUB) 978-3-11-061985-0

Volume 5
Systems Reliability Engineering: Modeling and Performance Improvement
Amit Kumar, Mangey Ram (Eds.)
ISBN 978-3-11-060454-2, e-ISBN (PDF) 978-3-11-061737-5
e-ISBN (EPUB) 978-3-11-061754-2

Volume 4
Systems Performance Modeling
Adarsh Anand, Mangey Ram (Eds.)
ISBN 978-3-11-060450-4, e-ISBN (PDF) 978-3-11-061905-8
e-ISBN (EPUB) 978-3-11-060763-5

Volume 3
Computational Intelligence: Theoretical Advances and Advanced Applications
Dinesh C. S. Bisht, Mangey Ram (Eds.)
ISBN 978-3-11-065524-7, e-ISBN (PDF) 978-3-11-067135-3
e-ISBN (EPUB) 978-3-11-066833-9

Volume 2
Supply Chain Sustainability: Modeling and Innovative Research Frameworks
Sachin Kumar Mangla, Mangey Ram (Eds.)
ISBN 978-3-11-062556-1, e-ISBN (PDF) 978-3-11-062859-3,
e-ISBN (EPUB) 978-3-11-062568-4

Volume 1
Soft Computing: Techniques in Engineering Sciences
Mangey Ram, Suraj B. Singh (Eds.)
ISBN 978-3-11-062560-8, e-ISBN (PDF) 978-3-11-062861-6,
e-ISBN (EPUB) 978-3-11-062571-4